Handbook of Mechanical Vibrations and Noise Engineering

Handbook of Mechanical Vibrations and Noise Engineering

Contributors

Quansheng Ji, Xiaomei Ji et al.

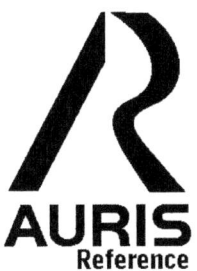

www.aurisreference.com

Handbook of Mechanical Vibrations and Noise Engineering

Contributors: Quansheng Ji, Xiaomei Ji et al.

Published by Auris Reference Limited
www.aurisreference.com

United Kingdom

Copyright 2016
Printed in 2017 for Sale in the Indian Subcontinent

The information in this book has been obtained from highly regarded resources. The copyrights for individual articles remain with the authors, as indicated. All chapters are distributed under the terms of the Creative Commons Attribution License, which permit unrestricted use, distribution, and reproduction in any medium, provided the original author and source are credited.

Notice

Contributors, whose names have been given on the book cover, are not associated with the Publisher. The editors and the Publisher have attempted to trace the copyright holders of all material reproduced in this publication and apologise to copyright holders if permission has not been obtained. If any copyright holder has not been acknowledged, please write to us so we may rectify.

Reasonable efforts have been made to publish reliable data. The views articulated in the chapters are those of the individual contributors, and not necessarily those of the editors or the Publisher. Editors and/or the Publisher are not responsible for the accuracy of the information in the published chapters or consequences from their use. The Publisher accepts no responsibility for any damage or grievance to individual(s) or property arising out of the use of any material(s), instruction(s), methods or thoughts in the book.

Handbook of Mechanical Vibrations and Noise Engineering

ISBN: 978-1-78154-966-7

British Library Cataloguing in Publication Data
A CIP record for this book is available from the British Library

Printed in the United Kingdom

Exclusively distributed by CBS Publishers & Distributors Pvt. Ltd.

Sales & Distribution Rights only for India, Pakistan, Bangladesh, Sri Lanka, Nepal and Bhutan. This book is not to be sold outside these territories.

Contents

List of Abbreviations .... *vii*

List of Contributors .... *ix*

Preface .... *xiii*

Chapter 1 **A New Differential Operator Method to Study the Mechanical Vibration** 1

Quansheng Ji, Xiaomei Ji, Linhong Ji, Yuxi Zheng

Chapter 2 **Characteristics of Mechanical Noise during Motion Control Applications** 11

Mehmet Emin Yüksekkaya

Chapter 3 **Theoretical Modelling and Effectiveness Study of Slotted Stand-Off Layer Damping Treatment for Rail Vibration and Noise Control** 27

Caiyou Zhao and Ping Wang

Chapter 4 **Study on Noise Prediction and Reduction in Coupled Workshops Using SEA Method** 57

Ye Lei, Jie Pan, and Meiping Sheng

Chapter 5 **Influence of Sound Vibration on Diamond-Like Carbon Deposition Rate** 77

Syed Md. Ihsanul Karim, Mohammad Asaduzzaman Chowdhury, and Md. Maksud Helali

Chapter 6 **A Noise Level Prediction Method Based on Electro-Mechanical Frequency Response Function for Capacitors** 95

Lingyu Zhu, Shengchang Ji, Qi Shen, Yuan Liu, Jinyu Li, Hao Liu

Chapter 7 **Experimental Investigations of Noise Control in Planetary Gear Set by Phasing** 117

S. H. Gawande and S. N. Shaikh

Chapter 8 **Noise Source Identification of a Ring-Plate Cycloid Reducer Based on Coherence Analysis** 145

Bing Yang and Yan Liu

Chapter 9 **Noise of Induction Machines** 157

Marcel Janda, Ondrej Vitek and Vitezslav Hajek

Chapter 10	**Automotive Applications of Active Vibration Control** 181
	Ferdinand Svaricek, Tobias Fueger, Hans-Juergen Karkosch, Peter Marienfeld and Christian Bohn
Chapter 11	**Progress and Recent Trends in the Torsional Vibration of Internal Combustion Engine** 203
	Liang Xingyu, Shu Gequn, Dong Lihui, Wang Bin and Yang Kang
Chapter 12	**Magnetic Levitation Technique for Active Vibration Control** 243
	Md. Emdadul Hoque and Takeshi Mizuno
	Citations 271
	Index 273

List of Abbreviations

CFT	Complex Fourier transform
CLF	Coupling Loss Factor
CVD	Chemical Vapor Deposition
DAQ	Data Acquisition
DLF	Damping Loss Factor
EDX	X-Ray Spectrometry
EMFRF	Electro Mechanical Frequency Response Function
HVDC	High Voltage Direct Current
IRFT	Inverse Real Fourier transform
LECD	Localized Electrochemical Deposition
PD	Proportional Derivative
PDV	Portable Digital Vibrometer
SEA	Statistical Energy Analysis
SIL	Sound Intensity
SPL	Sound Pressure Level
SWL	Sound Power Level
TAF	Torque Amplification Factor
XRD	X-Ray Diffraction

List of Contributors

Quansheng Ji
Shandong Iron and Steel Company Ltd., Jinan, China

Xiaomei Ji
Department of Mathematical Sciences, Yeshiva University, New York, USA

Linhong Ji
The State Key Laboratory of Tribology, Tsinghua University, Beijing, China

Yuxi Zheng
Department of Mathematical Sciences, Yeshiva University, New York, USA

Mehmet Emin Yüksekkaya
College of EngineeringUsak University, Turkey

Caiyou Zhao
Key Laboratory of High-Speed Railway Engineering, Ministry of Education, Southwest Jiaotong University, Chengdu 610031, China
School of Civil Engineering, Southwest Jiaotong University, Chengdu 610031, China

Ping Wang
Key Laboratory of High-Speed Railway Engineering, Ministry of Education, Southwest Jiaotong University, Chengdu 610031, China
School of Civil Engineering, Southwest Jiaotong University, Chengdu 610031, China

Ye Lei
School of Mechanical and Chemical Engineering, The University of Western Australia, Nedlands, Perth, WA 6009, Australia
School of Marine Engineering, Northwestern Polytechnical University, Xi'an 710072, China

Jie Pan
School of Mechanical and Chemical Engineering, The University of Western Australia, Nedlands, Perth, WA 6009, Australia

Meiping Sheng
School of Marine Engineering, Northwestern Polytechnical University, Xi'an 710072, China

Syed Md. Ihsanul Karim
Bangladesh Industrial Technical Assistance Centre (BITAC), Ministry of Industries, Dhaka 1208, Bangladesh

Mohammad Asaduzzaman Chowdhury
Department of Mechanical Engineering, Dhaka University of Engineering and Technology, Gazipur 1700, Bangladesh

Md. Maksud Helali
Department of Mechanical Engineering, Bangladesh University of Engineering and Technology, Dhaka 1000, Bangladesh

Lingyu Zhu
State Key Laboratory of Electrical Insulation and Power Equipment, Xi'an Jiaotong University, Xi'an, Shaanxi, China

Shengchang Ji
State Key Laboratory of Electrical Insulation and Power Equipment, Xi'an Jiaotong University, Xi'an, Shaanxi, China

Qi Shen
Shaoxing Electric Power Bureau, Shaoxing, Zhejiang, China

Yuan Liu
State Key Laboratory of Electrical Insulation and Power Equipment, Xi'an Jiaotong University, Xi'an, Shaanxi, China

Jinyu Li
State Key Laboratory of Electrical Insulation and Power Equipment, Xi'an Jiaotong University, Xi'an, Shaanxi, China

Hao Liu
State Key Laboratory of Electrical Insulation and Power Equipment, Xi'an Jiaotong University, Xi'an, Shaanxi, China

S. H. Gawande
Department of Mechanical Engineering, M. E. Society's College of Engineering, Pune, Maharashtra, India

S. N. Shaikh
Department of Mechanical Engineering, AISSMS College of Engineering, Pune, Maharashtra, India

Bing Yang
School of Mechanical Engineering, Dalian Jiaotong University, Dalian 116028, China

Yan Liu
School of Traffic and Transportation Engineering, Dalian Jiaotong University, Dalian 116028, China

Marcel Janda
Brno University of Technology, Czech Republic

Ondrej Vitek
Brno University of Technology, Czech Republic

Vitezslav Hajek
Brno University of Technology, Czech Republic

Ferdinand Svaricek
University of the German Armed Forces Munich

Tobias Fueger
University of the German Armed Forces Munich

Hans-Juergen Karkosch
ContiTech Vibration Control GmbH

Peter Marienfeld
ContiTech Vibration Control GmbH

Christian Bohn
Technical University Clausthal Germany

Liang Xingyu
State Key Laboratory of Engines, Tianjin University, 300072 P. R. China

Shu Gequn
State Key Laboratory of Engines, Tianjin University, 300072 P. R. China

Dong Lihui
State Key Laboratory of Engines, Tianjin University, 300072 P. R. China

Wang Bin
State Key Laboratory of Engines, Tianjin University, 300072 P. R. China

Yang Kang
State Key Laboratory of Engines, Tianjin University, 300072 P. R. China

Md. Emdadul Hoque
Saitama University Japan

Takeshi Mizuno
Saitama University Japan

Preface

The study of vibrations is concerned with the oscillating motion of elastic bodies and the force associated with them. Mechanical vibration is defined as the measurement of a periodic process of oscillations with respect to an equilibrium point. Handbook of Mechanical Vibrations and Noise Engineering provides essential concepts involving vibrational analysis, uncertainty modeling, and vibration control. It should also give a good fundamental basis in computational results, mathematical modeling and assessment in performance of different systems and system components. In first chapter, we propose a unified differential operator method to study mechanical vibrations, solving inhomogeneous linear ordinary differential equations with constant coefficients. In second chapter, an extensive analysis of the mechanical noise due to the building vibration has been analyzed and possible solutions to the problem discussed. Third chapter presents an exploratory study of the slotted stand-off layer damping treatment for rail vibration and noise control using both theoretical analysis and laboratory tests. A theoretical model for predicting noise reduction in coupled workshops is presented in fourth chapter by using statistical energy analysis (SEA) method. An opening between the coupled workshops is considered into the theoretical model properly. In fifth chapter, an attempt is made to investigate the effect of sound vibration, in particular, the frequency of vibration on the deposition rate. In addition to deposition rate, the quality of deposited coating is also investigated. In sixth chapter, a new noise level prediction method is proposed based on a frequency response function considering both electrical and mechanical characteristics of capacitors. In seventh chapter, several techniques have been proposed to reduce gear noise and vibrations in recent years. Noise source identification of a ring-plate cycloid reducer based on coherence analysis has been presented in eighth chapter. Ninth chapter emphasizes on noise of induction machines and automotive applications of active vibration control are presented in tenth chapter. Eleventh chapter mainly refers to the literatures on torsional vibration issue published in recent years, summarizes on the modeling of torsional vibration, corresponding analysis methods, appropriate measures and torsional vibration control, and points out the problems to be solved in the study and some new research directions. Last chapter presents an application of zero-power controlled magnetic levitation for active vibration control. Vibration isolation are strongly required in the field of high-resolution measurement and micromanufacturing, for instance, in the submicron semiconductor chip manufacturing, scanning probe microscopy, holographic interferometry, cofocal optical imaging, etc. to obtain precise and repeatable results.

Chapter 1

A NEW DIFFERENTIAL OPERATOR METHOD TO STUDY THE MECHANICAL VIBRATION

Quansheng Ji[1], Xiaomei Ji[2], Linhong Ji[3], Yuxi Zheng[4]
[1]Shandong Iron and Steel Company Ltd., Jinan, China
[2]Department of Mathematical Sciences, Yeshiva University, New York, USA
[3]The State Key Laboratory of Tribology, Tsinghua University, Beijing, China
[4]Department of Mathematical Sciences, Yeshiva University, New York, USA

ABSTRACT

In this paper, we propose a unified differential operator method to study mechanical vibrations, solving inhomogeneous linear ordinary differential equations with constant coefficients. The main advantage of this new method is that the differential operator D in the numerator of the fraction has no effect on input functions (i.e., the derivative operation is removed) because we take the fraction as a whole part in the partial fraction expansion. The method in various variants is widely implemented in related fields in mechanics and engineering. We also point out that the same mistakes in the differential operator method are found in the related references [1-4].

INTRODUCTION

A very serious equipment accident occurred in 1982 in a certain factory in the iron and steel industry in Jinan, China, where the safety pin of the middle plate of a three-roll mill for mechanical protection was not broken, but the main reducer (2800 kw, center distance 1900 mm) was so damaged that the whole gear produced crack with six connection bolts whose diameter is 64 mm of highspeed side pulled off, resulting in that the production had been halted for more than 20 days, and leading to huge economic losses. In the analysis of such serious equipment accident, we employed differential operator method in [3], Laplace transform and modal analysis method to calculate the natural frequency of torsional vibration and torque amplification factor for the main transmission system in medium plate rolling-mill. It was surprising that some mistakes were found in the solutions to the dynamic resonance of the above system through differential operator method, which was inconsistent with the correct results by Laplace transform and modal analysis method. Having compared with Laplace transform, we proposed a new differential operator method and solved the above same problem, obtaining the correct results completely consistent with modal analysis method and Laplace transform. We thus solved the disastrous equipment accident with little time. Here we illustrate the new differential operator method as follows: that differential operator D of numerator part in the fraction has no effect on input function because we do not need derivative operation working on input function. However, the fraction is taken as a whole part, using partial fraction expansion. Here we apply the related property of D operator in [5,6]. In [5], there is strict mathematical foundation for the partial fraction expansion of D operator. Finally, we obtain particular solution and general solution based on Cramer's rule as well as initial conditions. This method is suitable and powerful for solutions to different governing equations in many related vibration problems in the mechanics and engineering fields, showing more flexibility and superiority to other methods. It has not only enriched analytic methods in mechanical dynamics but also made contributions to the dynamic analysis in metallurgical equipments, where it was originally reported in [7]. Currently, in general teaching references in [1,2], for constant coefficients differential equations, the differential operator D in numerator part in its fraction affects input function through derivative operation and gives rise to errors herein.

THE NEW DIFFERENTIAL OPERATOR METHOD

Without loss of generality, we introduce the new differential operator method via the rectal serial five masses torsional vibration system, see **Figure 1**, which is simplified in **Figure 2** in the transmission system in the above original accident in 1982 we mentioned, assuming initial conditions are zero (i.e., we suppose initial displacement and initial velocity as zero, and as a general description for initial conditions, see [8]). We state the related parameters as follows:

1) Motor (2800 kW); 2) Gear coupling; 3) Flywheel; 4) Main reducer(center distance 1900 mm); 5) Safety pin coupling; 6) Herringbone gear stand; 7) Universal joint; 8) Top roller; 9) Medium roller; 10) Steel plate; 11) Lower roller.

$j_i (i = 1, 2, 3, 4, 5)$: Moment of inertia;

$k_{ij} (i,= 1, 2, 3, 4; j=2,3,4,5)$: Torsional elastic coefficient of every axial segment (rigidity);

φ_i (i = 1, 2, 3, 4, 5): Angular position;

$\ddot{\varphi}_i$ (i = 1, 2, 3, 4, 5): Angular acceleration;

M_{ij} (i = 1, 2, 3, 4; j = 2, 3, 4, 5): Torque of every axial segment;

M: Excitation torque;

D: Differential operator on behalf of d/dt;

Without the viscosity and damping, we set up the governing equations in the system:

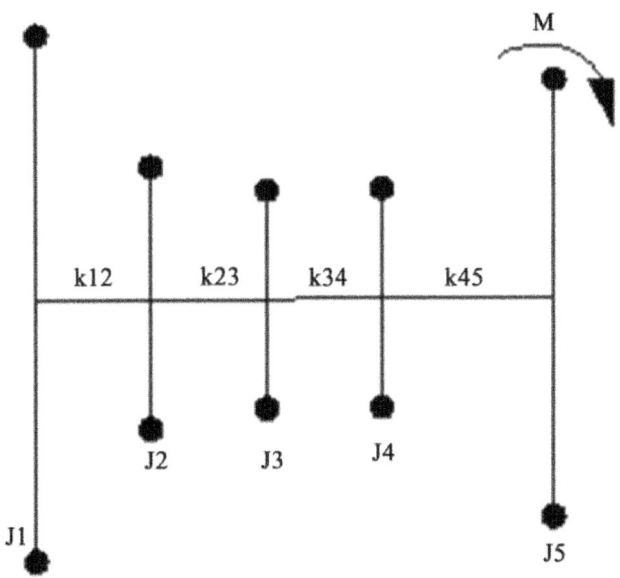

Figure 1. Diagram of five straight string masses system.

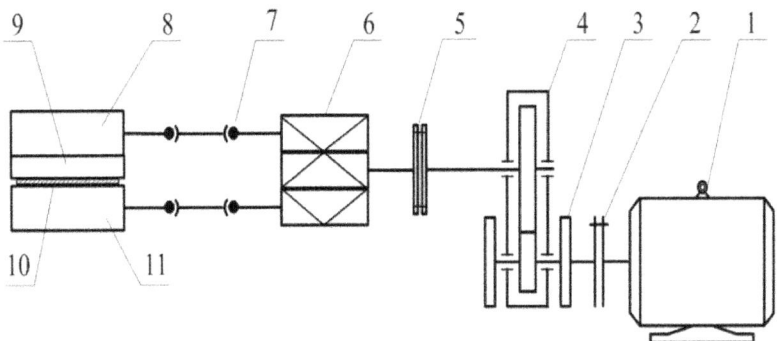

Figure 2. Diagram of transmission system.

$$\begin{cases} J_1\ddot{\phi}_1 + K_{12}(\phi_1 - \phi_2) = 0 \\ J_2\ddot{\phi}_2 + K_{23}(\phi_2 - \phi_3) - K_{12}(\phi_1 - \phi_2) = 0 \\ J_3\ddot{\phi}_3 + K_{34}(\phi_3 - \phi_4) - K_{23}(\phi_2 - \phi_3) = 0 \\ J_4\ddot{\phi}_4 + K_{45}(\phi_4 - \phi_5) - K_{34}(\phi_3 - \phi_4) = 0 \\ J_5\ddot{\phi}_5 - K_{45}(\phi_4 - \phi_5) = -M \end{cases} \quad (2.1)$$

Let $P_{ij}^2 = K_{ij} \dfrac{J_i + J_j}{J_i J_j}, i = 1,2,3,4; j = 2,3,4,5,$

and

$$\begin{cases} M_{12} = K_{12}(\phi_1 - \phi_2) \\ M_{23} = K_{23}(\phi_2 - \phi_3) \\ M_{34} = K_{34}(\phi_3 - \phi_4) \\ M_{45} = K_{45}(\phi_4 - \phi_5) \end{cases} \quad (2.2)$$

Then

$$\begin{cases} \ddot{M}_{12} + P_{12}^2 M_{12} - \dfrac{K_{12}}{J_2} M_{23} = 0 \\ \ddot{M}_{23} + P_{23}^2 M_{23} - \dfrac{K_{23}}{J_2} M_{12} - \dfrac{K_{23}}{J_3} M_{34} = 0 \\ \ddot{M}_{34} + P_{34}^2 M_{34} - \dfrac{K_{34}}{J_3} M_{23} - \dfrac{K_{34}}{J_4} M_{45} = 0 \\ \ddot{M}_{45} + P_{45}^2 M_{45} - \dfrac{K_{45}}{J_4} M_{34} = \dfrac{K_{45}}{J_5} M \end{cases} \quad (2.3)$$

And

$$\begin{cases} (D^2 + P_{12}^2) M_{12} - \dfrac{K_{12}}{J_2} M_{23} = 0 \\ (D^2 + P_{23}^2) M_{23} - \dfrac{K_{23}}{J_2} M_{12} - \dfrac{K_{23}}{J_3} M_{34} = 0 \\ (D^2 + P_{34}^2) M_{34} - \dfrac{K_{34}}{J_3} M_{23} - \dfrac{K_{34}}{J_4} M_{45} = 0 \\ (D^2 + P_{45}^2) M_{45} - \dfrac{K_{45}}{J_4} M_{34} = \dfrac{K_{45}}{J_5} M \end{cases} \quad (2.4)$$

Let the determinant of the coefficient of (2.4) be Δ. Then
$\Delta = (D^2 + P_1^2)(D^2 + P_2^2)(D^2 + P_3^2)(D^2 + P_4^2)$ where P_1, P_2, P_3, P_4 are from the first order to the fourth order natural frequencies in the main transmission system.

By Cramer's rule, we calculate $\Delta_{12}, \Delta_{23}, \Delta_{34}, \Delta_{45}$, where

$$\Delta_{12} = \frac{K_{12}K_{23}K_{34}K_{45}}{J_1 J_2 J_3 J_4 J_5} J_1 M,$$

$$\Delta_{23} = \frac{K_{12}K_{23}K_{34}K_{45}}{J_1 J_2 J_3 J_4 J_5}(J_1 + J_2)M + \frac{K_{23}K_{34}K_{45}}{J_3 J_4 J_5} D^2 M,$$

$$\Delta_{34} = \frac{K_{12}K_{23}K_{34}K_{45}}{J_1 J_2 J_3 J_4 J_5}(J_1 + J_2 + J_3)M$$
$$+ \frac{K_{34}K_{45}}{J_4 J_5}(P_{12}^2 + P_{23}^2)D^2 M + \frac{K_{34}K_{45}}{J_4 J_5} D^4 M.$$

and similarly, we obtain Δ_{45}.

We must pay attention to that we just take differential operator D as an algebra symbol, so it has no effect on input function, that is to say, we need not derivative operation, and we take the fraction with differential operator D in numerator and denominator as a whole part in partial fraction expansion. If we assume initial conditions of the system are zero and M is a step function, we can obtain analytic formulae of each axial segment torque individually below:

$$M_{12} = \frac{\Delta_{12}}{\Delta} = \frac{1}{(D^2 + P_1^2)(D^2 + P_2^2)(D^2 + P_3^2)(D^2 + P_4^2)} \frac{K_{12}K_{23}K_{34}K_{45}}{J_1 J_2 J_3 J_4 J_5} J_1 M$$

$$= \frac{J_1 M}{J_1 + J_2 + J_3 + J_4 + J_5} \times \left[\frac{P_2^2 P_3^2 P_4^2 (1-\cos P_1 t)}{(P_2^2 - P_1^2)(P_3^2 - P_1^2)(P_4^2 - P_1^2)} + \frac{P_1^2 P_3^2 P_4^2 (1-\cos P_2 t)}{(P_1^2 - P_2^2)(P_3^2 - P_2^2)(P_4^2 - P_2^2)} \right.$$
$$\left. + \frac{P_1^2 P_2^2 P_4^2 (1-\cos P_3 t)}{(P_1^2 - P_3^2)(P_2^2 - P_3^2)(P_4^2 - P_3^2)} + \frac{P_1^2 P_2^2 P_3^2 (1-\cos P_4 t)}{(P_1^2 - P_4^2)(P_3^2 - P_4^2)(P_2^2 - P_4^2)} \right];$$

$$M_{23} = \frac{(J_1 + J_2)}{J_1 + J_2 + J_3 + J_4 + J_5} M \left[\frac{P_2^2 P_3^2 P_4^2 (1-\cos P_1 t)}{(P_2^2 - P_1^2)(P_3^2 - P_1^2)(P_4^2 - P_1^2)} + \frac{P_1^2 P_3^2 P_4^2 (1-\cos P_2 t)}{(P_1^2 - P_2^2)(P_3^2 - P_2^2)(P_4^2 - P_2^2)} \right.$$
$$\left. + \frac{P_1^2 P_2^2 P_4^2 (1-\cos P_3 t)}{(P_1^2 - P_3^2)(P_2^2 - P_3^2)(P_4^2 - P_3^2)} + \frac{P_1^2 P_2^2 P_3^2 (1-\cos P_4 t)}{(P_1^2 - P_4^2)(P_3^2 - P_4^2)(P_2^2 - P_4^2)} \right]$$

$$- \frac{K_{23}K_{34}K_{45}}{J_3 J_4 J_5} M \left[\frac{1-\cos P_1 t}{(P_2^2 - P_1^2)(P_3^2 - P_1^2)(P_4^2 - P_1^2)} + \frac{1-\cos P_2 t}{(P_1^2 - P_2^2)(P_3^2 - P_2^2)(P_4^2 - P_2^2)} \right.$$
$$\left. + \frac{1-\cos P_3 t}{(P_1^2 - P_3^2)(P_2^2 - P_3^2)(P_4^2 - P_3^2)} + \frac{1-\cos P_4 t}{(P_1^2 - P_4^2)(P_3^2 - P_4^2)(P_2^2 - P_4^2)} \right];$$

$$M_{34} = \frac{(J_1+J_2+J_3)}{J_1+J_2+J_3+J_4+J_5}M\left[\frac{P_2^2 P_3^2 P_4^2(1-\cos P_1 t)}{(P_2^2-P_1^2)(P_3^2-P_1^2)(P_4^2-P_1^2)} + \frac{P_1^2 P_3^2 P_4^2(1-\cos P_2 t)}{(P_1^2-P_2^2)(P_3^2-P_2^2)(P_4^2-P_2^2)}\right.$$

$$\left.+\frac{P_1^2 P_2^2 P_4^2(1-\cos P_3 t)}{(P_1^2-P_3^2)(P_2^2-P_3^2)(P_4^2-P_3^2)} + \frac{P_1^2 P_2^2 P_3^2(1-\cos P_4 t)}{(P_1^2-P_4^2)(P_3^2-P_4^2)(P_2^2-P_4^2)}\right]$$

$$+\frac{K_{34}K_{45}}{J_4 J_5}M\left[\frac{P_1^2(1-\cos P_1 t)}{(P_2^2-P_1^2)(P_3^2-P_1^2)(P_4^2-P_1^2)} + \frac{P_2^2(1-\cos P_2 t)}{(P_1^2-P_2^2)(P_3^2-P_2^2)(P_4^2-P_2^2)}\right.$$

$$\left.+\frac{P_3^2(1-\cos P_3 t)}{(P_1^2-P_3^2)(P_2^2-P_3^2)(P_4^2-P_3^2)} + \frac{P_4^2(1-\cos P_4 t)}{(P_1^2-P_4^2)(P_3^2-P_4^2)(P_2^2-P_4^2)}\right]$$

$$-\frac{K_{34}K_{45}}{J_4 J_5}(P_{12}^2+P_{23}^2)M\left[\frac{1-\cos P_1 t}{(P_2^2-P_1^2)(P_3^2-P_1^2)(P_4^2-P_1^2)} + \frac{1-\cos P_2 t}{(P_1^2-P_2^2)(P_3^2-P_2^2)(P_4^2-P_2^2)}\right.$$

$$\left.+\frac{1-\cos P_3 t}{(P_1^2-P_3^2)(P_2^2-P_3^2)(P_4^2-P_3^2)} + \frac{1-\cos P_4 t}{(P_1^2-P_4^2)(P_3^2-P_4^2)(P_2^2-P_4^2)}\right];$$

$$M_{45} = \frac{(J_1+J_2+J_3+J_4)}{J_1+J_2+J_3+J_4+J_5}M\left[\frac{P_2^2 P_3^2 P_4^2(1-\cos P_1 t)}{(P_2^2-P_1^2)(P_3^2-P_1^2)(P_4^2-P_1^2)} + \frac{P_1^2 P_3^2 P_4^2(1-\cos P_2 t)}{(P_1^2-P_2^2)(P_3^2-P_2^2)(P_4^2-P_2^2)}\right.$$

$$\left.+\frac{P_1^2 P_2^2 P_4^2(1-\cos P_3 t)}{(P_1^2-P_3^2)(P_2^2-P_3^2)(P_4^2-P_3^2)} + \frac{P_1^2 P_2^2 P_3^2(1-\cos P_4 t)}{(P_1^2-P_4^2)(P_3^2-P_4^2)(P_2^2-P_4^2)}\right]$$

$$-\frac{K_{45}}{J_5}M\left[\frac{P_1^4(1-\cos P_1 t)}{(P_2^2-P_1^2)(P_3^2-P_1^2)(P_4^2-P_1^2)} + \frac{P_2^4(1-\cos P_2 t)}{(P_1^2-P_2^2)(P_3^2-P_2^2)(P_4^2-P_2^2)} + \frac{P_3^4(1-\cos P_3 t)}{(P_1^2-P_3^2)(P_2^2-P_3^2)(P_4^2-P_3^2)}\right.$$

$$\left.+\frac{P_4^4(1-\cos P_4 t)}{(P_1^2-P_4^2)(P_3^2-P_4^2)(P_2^2-P_4^2)}\right] + \frac{K_{45}}{J_5}(P_{12}^2+P_{23}^2+P_{34}^2)M$$

$$\times\left[\frac{P_1^2(1-\cos P_1 t)}{(P_2^2-P_1^2)(P_3^2-P_1^2)(P_4^2-P_1^2)} + \frac{P_2^2(1-\cos P_2 t)}{(P_1^2-P_2^2)(P_3^2-P_2^2)(P_4^2-P_2^2)} + \frac{P_3^2(1-\cos P_3 t)}{(P_1^2-P_3^2)(P_2^2-P_3^2)(P_4^2-P_3^2)}\right.$$

$$\left.+\frac{P_4^2(1-\cos P_4 t)}{(P_1^2-P_4^2)(P_3^2-P_4^2)(P_2^2-P_4^2)}\right] - \frac{K_{45}}{J_5}\left(P_{12}^2 P_{23}^2 + P_{12}^2 P_{34}^2 + P_{23}^2 P_{34}^2 - \frac{K_{23}K_{34}}{J_3^2} - \frac{K_{23}K_{34}}{J_2^2}\right)$$

$$\times M\left[\frac{1-\cos P_1 t}{(P_2^2-P_1^2)(P_3^2-P_1^2)(P_4^2-P_1^2)} + \frac{1-\cos P_2 t}{(P_1^2-P_2^2)(P_3^2-P_2^2)(P_4^2-P_2^2)}\right.$$

$$\left.+\frac{1-\cos P_3 t}{(P_1^2-P_3^2)(P_2^2-P_3^2)(P_4^2-P_3^2)} + \frac{1-\cos P_4 t}{(P_1^2-P_4^2)(P_3^2-P_4^2)(P_2^2-P_4^2)}\right].$$

If we follow the differential operator D of numerator part in its faction has effect on input function in general mathematics handbook [4] and the teaching material of constant differential equations in textbook [1,2] (that is to say, we carry out derivative operation), thus we can acquire the following items if M is a step function and initial conditions are zero:

$$\overline{\Delta_{12}} = \frac{K_{12}K_{23}K_{34}K_{45}}{J_1 J_2 J_3 J_4 J_5}J_1 M;$$

$$\overline{\Delta_{34}} = \frac{K_{12}K_{23}K_{34}K_{45}}{J_1 J_2 J_3 J_4 J_5}(J_1+J_2+J_3)M + \frac{K_{34}K_{45}}{J_4 J_5}(P_{12}^2+P_{23}^2)D^2 M + \frac{K_{34}K_{45}}{J_4 J_5}D^4 M = \frac{K_{12}K_{23}K_{34}K_{45}}{J_1 J_2 J_3 J_4 J_5}(J_1+J_2+J_3)M;$$

$$\overline{\Delta_{23}} = \frac{K_{12}K_{23}K_{34}K_{45}}{J_1J_2J_3J_4J_5}(J_1+J_2)M + \frac{K_{23}K_{34}K_{45}}{J_3J_4J_5}D^2M = \frac{K_{12}K_{23}K_{34}K_{45}}{J_1J_2J_3J_4J_5}(J_1+J_2)M;$$

$$\overline{\Delta_{45}} = \frac{K_{12}K_{23}K_{34}K_{45}}{J_1J_2J_3J_4J_5}(J_1+J_2+J_3+J_4)M - \frac{K_{45}}{J_5}D^6M - \frac{K_{45}}{J_5}\left(P_{12}^2+P_{23}^2+P_{34}^2\right)D^4M - \frac{K_{45}}{J_5}$$

$$\times \left[P_{12}^2P_{23}^2 + P_{12}^2P_{34}^2 + P_{23}^2P_{34}^2 - \frac{K_{23}K_{34}}{J_3^2} - \frac{K_{23}K_{34}}{J_2^2}\right]D^2M = \frac{K_{12}K_{23}K_{34}K_{45}}{J_1J_2J_3J_4J_5}(J_1+J_2+J_3+J_4)M;$$

$$\overline{M_{12}} = \frac{J_1}{J_1+J_2+J_3+J_4+J_5}M\left[\frac{P_2^2P_3^2P_4^2(1-\cos P_1t)}{(P_2^2-P_1^2)(P_3^2-P_1^2)(P_4^2-P_1^2)} + \frac{P_1^2P_3^2P_4^2(1-\cos P_2t)}{(P_1^2-P_2^2)(P_3^2-P_2^2)(P_4^2-P_2^2)}\right.$$

$$\left. + \frac{P_1^2P_2^2P_4^2(1-\cos P_3t)}{(P_1^2-P_3^2)(P_2^2-P_3^2)(P_4^2-P_3^2)} + \frac{P_1^2P_2^2P_3^2(1-\cos P_4t)}{(P_1^2-P_4^2)(P_3^2-P_4^2)(P_2^2-P_4^2)}\right];$$

$$\overline{M_{23}} = \frac{(J_1+J_2)}{J_1+J_2+J_3+J_4+J_5}M\left[\frac{P_2^2P_3^2P_4^2(1-\cos P_1t)}{(P_2^2-P_1^2)(P_3^2-P_1^2)(P_4^2-P_1^2)} + \frac{P_1^2P_3^2P_4^2(1-\cos P_2t)}{(P_1^2-P_2^2)(P_3^2-P_2^2)(P_4^2-P_2^2)}\right.$$

$$\left. + \frac{P_1^2P_2^2P_4^2(1-\cos P_3t)}{(P_1^2-P_3^2)(P_2^2-P_3^2)(P_4^2-P_3^2)} + \frac{P_1^2P_2^2P_3^2(1-\cos P_4t)}{(P_1^2-P_4^2)(P_3^2-P_4^2)(P_2^2-P_4^2)}\right];$$

$$\overline{M_{34}} = \frac{(J_1+J_2+J_3)}{J_1+J_2+J_3+J_4+J_5}M\left[\frac{P_2^2P_3^2P_4^2(1-\cos P_1t)}{(P_2^2-P_1^2)(P_3^2-P_1^2)(P_4^2-P_1^2)} + \frac{P_1^2P_3^2P_4^2(1-\cos P_2t)}{(P_1^2-P_2^2)(P_3^2-P_2^2)(P_4^2-P_2^2)}\right.$$

$$\left. + \frac{P_1^2P_2^2P_4^2(1-\cos P_3t)}{(P_1^2-P_3^2)(P_2^2-P_3^2)(P_4^2-P_3^2)} + \frac{P_1^2P_2^2P_3^2(1-\cos P_4t)}{(P_1^2-P_4^2)(P_3^2-P_4^2)(P_2^2-P_4^2)}\right];$$

$$\overline{M_{45}} = \frac{(J_1+J_2+J_3+J_4)}{J_1+J_2+J_3+J_4+J_5}M\left[\frac{P_2^2P_3^2P_4^2(1-\cos P_1t)}{(P_2^2-P_1^2)(P_3^2-P_1^2)(P_4^2-P_1^2)} + \frac{P_1^2P_3^2P_4^2(1-\cos P_2t)}{(P_1^2-P_2^2)(P_3^2-P_2^2)(P_4^2-P_2^2)}\right.$$

$$\left. + \frac{P_1^2P_2^2P_4^2(1-\cos P_3t)}{(P_1^2-P_3^2)(P_2^2-P_3^2)(P_4^2-P_3^2)} + \frac{P_1^2P_2^2P_3^2(1-\cos P_4t)}{(P_1^2-P_4^2)(P_3^2-P_4^2)0}\right](P_2^2-P_4^2).$$

Obviously, the above result of torsional vibration dynamic resonance in every axial segment shows only related to the distribution coefficients of moment of inertia with the same coefficients of frequency difference and the same time of every axial segment attained by peak moment, but it is inappropriate in practice. Above result seems only like the ordinary dynamic questions (the starting of acceleration movement and the braking deceleration movement); however, it apparently does not belong to questions of torsional vibration. Consequently, the results via the new differential operator method we mentioned above are not only as same as that by Laplace transform and model analysis method, but also coincide with that of recently widely used in modal matrix method in mechanical vibration.

Now we take main transmission system in 2300 medium plate rolling-mill in the original serious accident we referred as an example, although it is some roughly approximate to the rectal serial five masses system (generally speaking, it should be taken as a branch system, and here it is only for a practical application of above method).

We solve the quantities as follows by the related data:

$P_1 = 182.55, P_2 = 349.9,$
$P_3 = 443.79, P_4 = 741.71.$

Then we have

$M_{12} = (0.0859 - 0.1511 \cos P_1 t + 0.1085 \cos P_2 t$
$\qquad - 0.0442 \cos P_3 t + 0.009 \cos P_4 t) M;$

$M_{23} = (1.0179 - 1.297 \cos P_1 t + 0.001 \cos P_2 t$
$\qquad + 0.3115 \cos P_3 t - 0.03548 \cos P_4 t) M;$

$M_{34} = (0.99483 - 1.0354 \cos P_1 t - 0.000437 \cos P_2 t$
$\qquad - 0.0347 \cos P_3 t - 0.0749 \cos P_4 t) M;$

$M_{45} = (0.93943 - 0.7955 \cos P_1 t - 0.00628 \cos P_2 t$
$\qquad -0.1742 \cos P_3 t - 0.0338 \cos P_4 t) M$

According to appearing time of the peak moment, we obtain the peak moment value in the **Table 1** by the formula below:

Torque Amplification Factor (TAF) = peak moment value/Applied to excited vibration moment maximum value on roller.

CONCLUSIONS

From the above analysis we have the same solutions as that in Laplace transformation and modal analysis method, such that we solved the accident quickly. According to the torque amplification factor (TAF) in our analysis, the related unit in the factory revised and adjusted the size of the safety pin quickly, and thereafter such similar huge accident has never happened.

As a problem in dynamic resonance of torsional vibration, model analysis method belongs to numerical computing method, but it cannot analyze the effects on all parameters in practical applications. It is easy for us to utilize the new differential operator method to analyze the effect on each parameter compared with other methods. It has been widely extended and applied to all related fields in mechanics and engineering, which we clearly clarify the errors mentioned in the references [1-4] here. In addition, the new differential operator method has been successfully implemented to solve other governing equations in the problems in mechanical dynamics, see pp. 71-89 in Chapter 15 in [9], where [9] wins the second award of Scientific and Technical Prize in Chinese Mechanical Industry in 2010.

Table 1. Torsional vibration quantities in the example in this paper

Axial segment	Peak moment appearing time	Peak moment value	(TAF)
M_{12}	0.018 s	$0.35\,M$	0.35
M_{23}	0.156 s	$2.636\,M$	2.636
M_{34}	0.12 s	$2.096\,M$	2.096
M_{45}	0.19 s	$1.854\,M$	1.854

ACKNOWLEDGEMENTS

We very much appreciate that Professors of Xianzhi Liu in Shandong University and Weichang Qian in Shanghai University who have supported this paper greatly.

REFERENCES

1. Department of Mathematics, Shanghai Tongji University, "Advanced Mathematics I," 6th Edition (in Chinese), Publishing House of High Education, Beijing, 2007, pp. 350-352.
2. C. H. Edwards and D. E. Penney, "Elementary Differential Equations," 6th Edition, Pearson Prentice Hall, Upper Saddle River, 2008, pp. 340-345.
3. S. E. Kerufunikefu, "Mechanical Dynamics with Elastic Key Ring," (in Russian, No Further Version), Kiev, 1961, pp. 86-89.
4. "Mathematics Handbook," (in Chinese), Publishing House of People Education, Beijing, 1979, pp. 675-677 (reprinted by Publishing House of High Education, Beijing, 1999).

5. R. P. Agnew, "Differential Equations," 2nd Edition, McGraw-Hill, Book Company, Inc., New York, Toronto, London, 1960, pp. 216-250.
6. L. E. Ėl'sgolts, "Differential Equations," in the Series International Monographs on Advanced Mathematics and Physics, Gordon and Breach Publishers, Inc., New York, and Hindustan Publishing Corporation, Delhi, 1961, pp. 145-155.
7. C.-S. Ji, "A New Solution of Constant Differential Equation Group by Differential Operator and Application in Calculating Rolling-Mill Torsion Vibration," Proceedings of the 6th International Modal Analysis Conference, Kissimmee, 1-4 February 1988, pp. 598-602.
8. Z. Zheng, "Mechanical Vibration I," (in Chinese), Publishing House of Mechanical Industry, Beijing, 1980.
9. "Mechanical Design Handbook," (in Chinese), Publishing House of Chemical Industry, Beijing, 2009.

Chapter 2

CHARACTERISTICS OF MECHANICAL NOISE DURING MOTION CONTROL APPLICATIONS

Mehmet Emin Yüksekkaya
College of EngineeringUsak University, Turkey

INTRODUCTION

Signal characteristics and processing are an important factor during today's digital world including the motion control and strain measurement applications. A digital signal is someone that can assume only a finite set of values is given for both the dependent and independent variables being analyzed (Smith, 2006). The independent variables are usually time or space; and the dependent variables are usually amplitudes. To use digital signal processing tools effectively, an analog signal must be converted into its digital representation in time space. In practice, this is implemented by using an analog-to-digital converter (A/D), which is an integral part of data acquisition (DAQ) cards (Vaseqhi, 2009). One of the most important parameters of an analog input system is the sampling rate at which the DAQ card samples an incoming signal. During the measurement and processing of the signal digitally, it is common to face noise problems interfacing the signals captured (Yuksekkaya, 1999; Chu & George, 1999; Kester, 2004). The noise could be coming from various sources with different characteristics and affecting the measurement systems. Once the signal is contaminated with the noise, the reading from the instruments will not be representing the actual situation of the physical phenomenon being captured. Therefore, it is an important area of practice to analyze the characteristics of the noise for any implementation of the signal analysis before constructing further refinements for data analysis. Furthermore, it would be a more practical to take some precautions in order to reduce the effects of the noise on the signal. Even it is possible to use some tools to decrease the effects of the noise to the signal ratio, it would be more practical to eliminate the noise as much as possible at the first hand. It is also evident from the industrial applications that the cost of initial investments for any noise elimination applications is cheaper than that of later investments.

Computer-based data acquisition systems using small computers have been successfully applied in many industrial applications including the motion control processes producing high performances at relatively low costs. As the investment cost of data processing systems decreases, it is getting more common to see a number of data acquisition systems implemented applications in our daily life. The benefits of a data acquisition system include: an improved analysis, accuracy and consistency, reduced analysis time and cost, and lower response time for an out-of-control situation regarding quality. It could be easily noticed that there would be a tremendous amount of noise superimposed on the signal coming from the measurement units. The noise could be coming from different sources depending on the application area. The main sources of the noise, however, are mechanical and electrical noises commonly found at the industrial applications (Yuksekkaya, 1999). Therefore, refinements are necessary for most of the times so that the noise problems could be eliminated from the signal in order to make an accurate measurement during the motion control.

During the industrial applications such as CNC controlled lathes and load cells taking the dynamic measurements, a considerable amount of mechanical noise could be superimposed to the signal from the ground due to the vibration of the buildings. The mechanical noise problem could damage the reading from the instruments due to the noise superimposed to the signal. In this text, an extensive analysis of the mechanical noise due to the building vibration has been analyzed and possible solutions to the problem discussed.

DIAGNOSTICS OF NOISE IN THE SIGNAL

As stated, digital signal is a finite set of values in both the dependent and independent variables. One of the most important parameters of an analog input system is the rate at which the DAQ card samples an incoming signal. A fast sampling rate acquires more points in a given time. As a result, a better representation of the signal is formed. Sampling too slow may result in a poor representation of the signal. This may cause a misrepresentation of a signal, which is commonly known as an aliasing effect. In order to avoid aliasing effects, the *Nyquist Theorem* states that a signal must be acquired at the rate greater than twice the maximum frequency component in the signal acquired (Ramirez, 1985). Figure 1 indicates the basic divisions of different signal types. The most fundamental division is stationary and non-stationary signals. Stationary signals are characterized by average properties that do not vary with time and independent of the particular sample record used to determine them.

The term "non-stationary" covers all signals that do not satisfy the requirements for stationary signals. Computer-based data acquisition systems using small computers have been successfully applied in many applications producing high performances at relatively low costs. The benefits of a data acquisition system include: an improved analysis, accuracy and consistency, reduced analysis time and cost, and lower response time for an out-of-control situation regarding quality. A typical data acquisition system consists of several parts: a signal conditioning module, a data acquisition hardware (A/D converter), analysis hardware, and data analysis software.

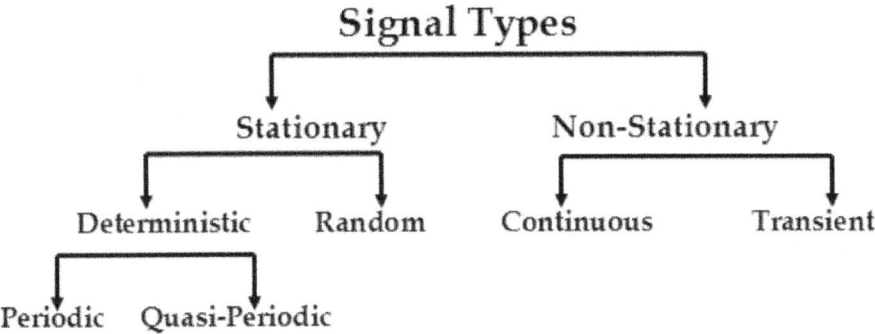

Figure 1. Classification of signal types.

In any digitally working environment, most of the time, there were two types of main noise problems that needed to be solved in order to have a meaningful result as follows (Yuksekkaya & Oxenham 1999; Yuksekkaya et. al. 2008):

1. Electrical Noise:
    i. Static
        i. Fluorescent lamps
        ii. Computer screens
        iii. Others
    ii. Dynamic
        i. AC power lines
        ii. Stepper motor
        iii. Transformers
        iv. Magnetic fields from other equipment

2. Mechanical Noise:
    i. Vibration from stepper motor
    ii. Vibration from building
    iii. Vibration from the other sources

Electrical Noise

Tensile testing devices are a combination of a strain measurement unit and a stepper motor which drives the measurement unit. During the processing of a strain measurement, the location of measurement unit should be precisely located in order to have an accurate stress-stain reading from the instrument. Most of the time, strain measurements are taken in the presence of electrical and magnetic fields, which can superimpose electrical noise on the measurement signals. If the electrical noise is not controlled properly, the noise can lead to inaccurate results and incorrect interpretation of the signals coming from the strain gages as well as the inaccurate location data. In order to control the noise level and maximize the signal-to-noise ratio, it is first necessary to understand the types and characteristics of electrical noise as well as the sources of such noises. Without understanding the noise and its sources, it is impossible to apply the most effective noise-reduction methods on any particular instrumentation problem.

Virtually, every electrical device that generates, consumes, or transmits power is a potential source for causing noise in strain gages. In general, the higher the voltage or current level and the closer the circuit is to the electrical device, the greater the induced noise will be superimposed to the signal. A list of common sources of electrical noise could be found in any signal analysis textbooks and electrical noise from those sources could be categorized into two basic types, that is: electrostatic and magnetic noises (Croft et al., 2006; Agres, 2007). The characteristics of these two types of noise are different; and they require different noise reduction techniques in order to eliminate their effects on the signals. Most of the noise coming from outside may be eliminated by using shielded, twisted cables and eliminating the ground loops in the system (more than one connection of the system to the ground). Furthermore, electromagnetic noises could be eliminated by using a special designed apparatus named as *Faraday Cage* if the application requires it.

Mechanical Noise

It would be practically possible to see that a strain measurement signal could be so sensitive that it would be continuously picking up mechanical noise from different sources such as from the buildings and from the stepper motors. It is necessary to analyze the building and stepper motor vibration sources separately. An extensive analysis of the mechanical noise coming from the building and potential solutions for the vibration sources are given as follows:

Building Vibration

Laboratory measurement instruments could be located in either stationary or mobile laboratories depending on the type of measurements necessary to perform. Regardless of the location of the instruments, it is a known fact that the ground vibration will affect the instrument's reading if correct precautions will not be applied. A considerable amount of mechanical noise would be coming from the ground due to the vibration of the building. In order to eliminate, or at least minimize, the effect of the ground movement as much as possible, usually the testing instrument was mounted on the top of a heavy marble block that was supported by a spring-like material. In order to analyze the vibrating mechanical system, let us consider an object hanging from a spring as shown in Figure 2 (Halliday et al., 2007; Zill & Cullen, 2006).

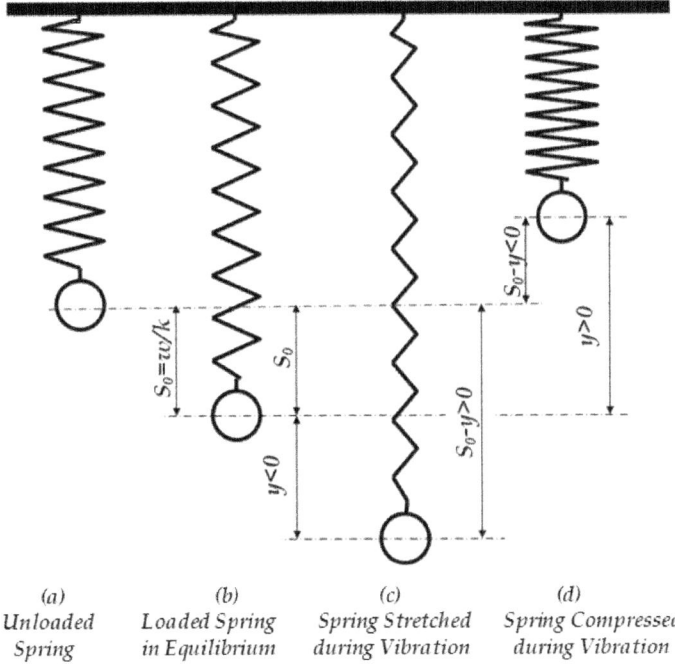

(a) Unloaded Spring
(b) Loaded Spring in Equilibrium
(c) Spring Stretched during Vibration
(d) Spring Compressed during Vibration

Figure 2. Schematic of mass spring system.

The weight, w, of the object is the magnitude of the force of gravity acting on the measurement instrument. The mass, m, of the object is related to the weight of the object, w as follows:

$$w = mg \qquad (1)$$

where g is the acceleration of the gravity. In order to keep complications to a minimum for the sake of analysis purposes, it was assumed that the spring obeys Hooke's law, that is, force is proportional to displacement. Namely,

$$f_s = ky \qquad (2)$$

Where y is the displacement and k is the stiffness of the spring element. A realistic analysis of the vertical motion of the mass would take into account not only the elastic and gravitational forces, but also, the effects of the friction affecting the system and all other forces that act externally on the suspended mass. Considering the other forces diminishes the amplitude of the vibration. In order to keep the analysis as simple as possible, let us do not take them into the account now and talk about the details later.

Observing that $m = w/g$ and applying Newton's second law of motion, (*force = mass * acceleration*) to the system gives a very popular equation:

$$\frac{w}{g}\frac{d^2y}{dt^2} + ky = 0 \qquad (3)$$

This is an equation of harmonic motion and its solution was discussed in almost every differential equation book (Halliday et al., 2007; Zill & Cullen, 2006). The solution for such a system is:

$$y = A\cos\sqrt{\frac{kg}{w}}t + B\sin\sqrt{\frac{kg}{w}}t \qquad (4)$$

If the term $\sqrt{\frac{kg}{w}}$ is set to be ω, then Equation 4 can be written by using a periodic function, as used above, in order to have a more compact form as follows:

$$y = A\cos\omega t + B\sin\omega t \qquad (5)$$

Regardless of the values of A and B, that is, regardless of how the system is set in motion, Equation 4 describes the periodic motion with the period of

$$2\pi\sqrt{\frac{w}{kg}} \qquad (6)$$

or frequency of

$$\frac{1}{2\pi}\sqrt{\frac{kg}{w}} \qquad (7)$$

Whether there is friction in the system or not, the quantity $\frac{1}{2\pi}\sqrt{\frac{kg}{w}}$ is called the *natural frequency*, (ω_n), of the system because this is the frequency at which the spring-mass system would vibrate naturally if no frictional or non-elastic forces other than gravity were present in a given system.

As mentioned earlier, the process by which free vibration diminishes in the amplitude is called the damping effect. If the damping effect presents in the system, the energy of the vibrating system will be dissipated by various mechanisms affecting the system, and often, more than one mechanism could be present in the system at the same time. In such a system, the damping force, f_D, is related to the velocity across the linear viscous damper by the following equation:

$$f_D = c\frac{dy}{dt} \qquad (8)$$

where the constant, c, is the viscous damping coefficient and has the unit of force*time/lengthforce*time/length. It is important to get the correct damping factor for a given system. Unfortunately, unlike the stiffness of the spring, the damping coefficient cannot be calculated from the dimensions of the structure or the size of the structural elements. Therefore, it should be evaluated from the vibration experiments on actual structures in order to get a precise coefficient of damping ratio for minimizing the effects of mechanical vibration to the measurement instruments.

This system is usually called a mass-spring-damp system, and its governing equation can be written according to the Newton's second law of motion as follows:

$$\frac{w}{g}\frac{d^2y}{dt^2} + c\frac{dy}{dt} ky = p(t) \qquad (9)$$

where $p(t) = p\ 0\ sin(\omega t)$ is the function of the external force (from ground vibration, etc.) acting to the system. The nature of the free motion of the system will depend on the roots of the related characteristic equation of the second order differential equation given in the equation. The characteristic equation for this second order differential equation is given as follows:

$$-\frac{cg}{2w} \mp \frac{g}{2w}\sqrt{c^2 - \frac{4kw}{g}} \tag{10}$$

It is clear that g, k, and w are all positive quantities and c is a non-negative and real number. Therefore, the characteristic of the solution of this second order differential equation depends upon the term $\sqrt{c^2 - \frac{4kw}{g}}$.

It is clear that there are three possibilities depending upon the values of k, w, and c, for the solution of this second order differential equation namely,

$$\sqrt{c^2 - \frac{4kw}{g}} \begin{cases} > 0 \\ = 0 \\ < 0 \end{cases} \tag{11}$$

If $\sqrt{c^2 - \frac{4kw}{g}} > 0$, there is a relatively large amount of friction, and, naturally enough, the system or its motion is said to be over-damped. In this case, the roots of the characteristic equation are real and unequal. The general solution is given by:

$$y = Ae^{m_1 t} + Be^{m_2 t} \tag{12}$$

where both of the roots of the second order differential equation, m_1 and m_2, are negative. Thus, y approaches zero as time increases indefinitely. If $\sqrt{c^2 - \frac{4kw}{g}} = 0$, it is at the borderline in which the roots of the characteristic equation are equal and real. In this case, the free motion can be expressed as follows:

$$y = Ae^{mt} + Bte^{mt} = (A + Bt)e^{mt} \tag{13}$$

From the equation $\sqrt{c^2 - \frac{4kw}{g}}$, critical damping can be defined as:

$$c_{cr} = \frac{2w}{g}\omega_n = 2m\omega_n \tag{14}$$

Then, the damping ratio, ζ, is defined as:

$$\zeta = \frac{c}{2m\omega_n} = \frac{c}{c_{cr}} \tag{15}$$

If $\sqrt{c^2 - \frac{4kw}{g}} < 0$, the motion is said to be under-damped. The roots of the characteristic equation are then the conjugate complex numbers given by:

$$m_1, m_2 = -\frac{cg}{2w} \mp i\sqrt{\frac{kg}{w} - \frac{c^2g^2}{4w^2}} \tag{16}$$

where $i = \sqrt{-1}$. Then, the general solution for the differential equation is given by:

$$y = Ae^{-\frac{cg}{2w}t}\cos\sqrt{\frac{kg}{w} - \frac{c^2g^2}{4w^2}}t + Be^{-\frac{cg}{2w}t}\sin\sqrt{\frac{kg}{w} - \frac{c^2g^2}{4w^2}}t \tag{17}$$

By defining some of the terms in the equation given above differently and using some trigonometric identities, the solution of the equation may be written in a more compact form as follows:

$$\sqrt{\frac{kg}{w} - \frac{c^2g^2}{4w^2}} = \omega_n\sqrt{1-\zeta^2} = \omega_D \tag{18}$$

The differential equation is solved, subject to the initial conditions $y = y(0)$, and $\frac{dy}{dt} = dy/dt|_{y=0}$. The particular solution of such a system is given by:

$$y_p = \frac{p_0}{k\left[1-\frac{\omega^2}{\omega_n^2}\right]^2 + \left[2\zeta\left(\frac{\omega}{\omega_n}\right)\right]^2}\left[\left(1-\left(\frac{\omega}{\omega_n}\right)^2\right)\cos(\omega t) - \frac{2\zeta\omega}{\omega_n}\sin(\omega t)\right]$$
(19)

By setting

$$C = \frac{p_0\left(1-\frac{\omega^2}{\omega_n^2}\right)}{k\left[1-\frac{\omega^2}{\omega_n^2}\right]^2 + \left[2\zeta\left(\frac{\omega}{\omega_n}\right)\right]^2} \text{ and } D = \frac{-2p_0\zeta\frac{\omega}{\omega_n}}{k\left[1-\frac{\omega^2}{\omega_n^2}\right]^2 + \left[2\zeta\left(\frac{\omega}{\omega_n}\right)\right]^2},$$

the Equation 19 can be written as

$$y_p = C\sin(\omega t) + D\cos(\omega t) \tag{20}$$

However, the complete solution of Equation 9 consists of transient and steady parts given as follows:

$$y(t) = e^{-\zeta\omega_n t}\underbrace{(A\cos\omega_D + B\sin\omega_D}_{\text{transient}} + \underbrace{C\sin\omega t + D\cos\omega t}_{\text{steady state}} \tag{21}$$

where the constant A and B can be determined in terms of the initial displacement and initial velocity. The steady state deformation of the system, due to harmonic force given in Equation 19 can be rewritten as follows:

$$y(t) = y_0 \sin(\omega t - \phi) = \frac{p_0}{k} R_d \sin(\omega t - \phi) \tag{22}$$

where $y_0 = \sqrt{C^2 + D^2}$ and $\phi = \tan^{-1}\left(\frac{D}{C}\right)$. Substituting for C and D gives deformation response factor, R_d, and phase angle, f.

$$R_d = \frac{1}{\sqrt{\left[1-\frac{\omega^2}{\omega_n^2}\right]^2 + \left[2\zeta\left(\frac{\omega}{\omega_n}\right)\right]^2}} \tag{23}$$

and

$$\phi = \tan^{-1}\left(\frac{2\zeta\frac{\omega}{\omega_n}}{1-\frac{\omega^2}{\omega_n^2}}\right) \tag{24}$$

Differentiating Equation 22 gives the velocity as follows:

$$\frac{dy(t)}{dt} = \frac{p_0}{\sqrt{km}} R_v \cos(\omega t - \phi) \tag{25}$$

where R_v is the velocity response factor and related to R_d by

$$R_v = \frac{\omega}{\omega_n} R_d \tag{26}$$

After applying the basic definitions and solutions of the differential equation for a mass-spring-damper system, force transmission and vibration isolation can be taken into account as follows: Consider the mass-spring-damper system (The system is the instrument itself and any foundation making the total weight higher), shown in Figure 3 subjected to a harmonic force. The force transmitted to the base is given by:

$$f_T = f_s + f_D = ky(t) + c\frac{dy(t)}{dt} \tag{27}$$

Substituting Equation 22 for $y(t)$ and Equation 25 for $\frac{dy(t)}{dt}$, and using Equation 26 give:

$$f_T(t) = (y_{st})_0 R_d [k \sin(\omega t - \phi) + c\omega \cos(\omega t - \phi)] \tag{28}$$

The maximum value of $fT(t)$ over t is:

$$(f_T)_0 = (y_{st})_0 R_d \sqrt{k^2 + c^2\omega^2} \tag{29}$$

which after using $(y_{st})_0 = \frac{p_0}{k}$ and $\zeta = \frac{c}{2m\omega_n}$ can be expressed as:

$$\frac{(f_T)_0}{p_0} = R_d\sqrt{1+\left(2\zeta\frac{\omega}{\omega_n}\right)^2} \tag{30}$$

Substituting Equation 23 gives an equation for the ratio of the maximum transmitted force to the amplitude p_0 of the applied force, known as the transmissibility (*TR*) of the system for a mass-spring-damper application:

$$TR = \sqrt{\frac{1+\left(2\zeta\frac{\omega}{\omega_n}\right)^2}{\left[1-\frac{\omega^2}{\omega_n^2}\right]^2+\left[2\zeta\left(\frac{\omega}{\omega_n}\right)\right]^2}} \tag{31}$$

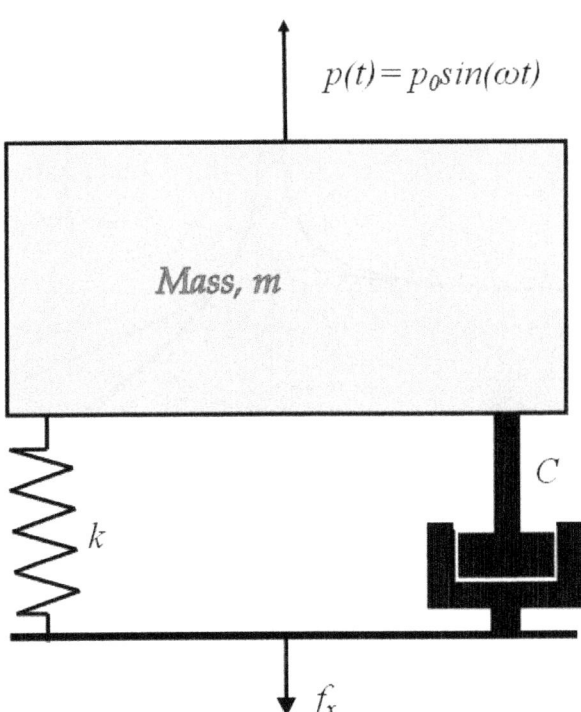

Figure 3. Simple spring-mass-damper system.

The transmissibility is plotted in Figure 4 as a function of the frequency ratio ω/ω_n, for several values of the damping ratio, ζ. For the transmitted force to be less than the applied force, the stiffness of the support system, and hence, the natural frequency, should be small enough so that the ratio of ω/ω_n should be bigger than $\sqrt{2}$ as seen in Figure 4. No damping is desired in the support system because, in this frequency range, damping increases the transmitted force. This implies a trade-off between a soft spring material to reduce the transmitted force and an acceptable static displacement. If the excitation frequency, ω, is much smaller than the natural frequency, ω_n, of the system, (*i.e.:* the mass is static while the ground beneath it is dynamic). This is the concept underlying isolation of a mass from a moving base by using a very flexible support system. For example, instruments or even buildings have been mounted on natural rubber bearings in order to isolate them from the ground-borne, vertical vibration (typically, with frequencies that range from 25 to 50 Hz) due to the rail traffic (Bozorgnia & Bertero, 2004; Chen & Lui, 2005). It would be also advisable to use rubber like material on the testing instruments in order to diminish the effects of the ground vibrations.

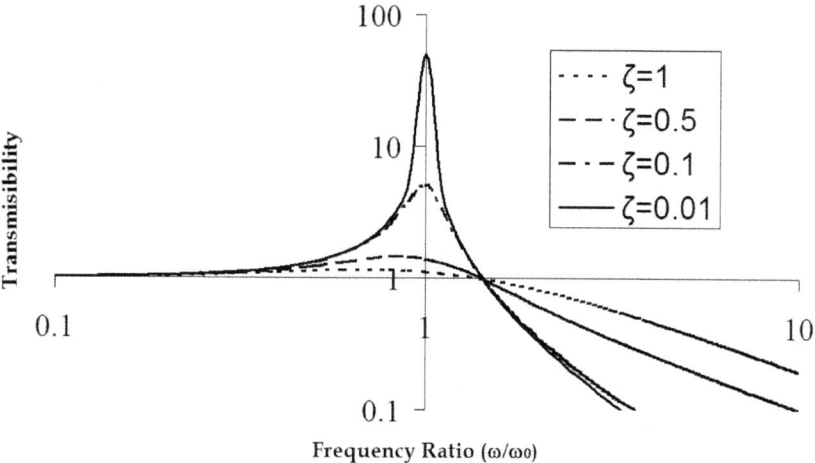

Figure 4. Transmissibility for harmonic excitation for various damping factors (Axes are in logarithmic scale).

Figure 5 shows a typical building vibration effect on the acquired signal from an instrument. In order to reduce the amount of vibration that is transmitted to the instrument, natural rubber-like materials, such as tennis balls, are the appropriate choice. Figure 6 shows the effect of vibration dampers on the instrument. As seen from the graph, the usage of vibration damper reduces the effects of mechanical vibrations significantly. Some further improvements can be achieved by precisely calculating (and if necessary modifying) the stiffness of the insulation material taking into consideration the low damping coefficient or increasing the weight of instrument.

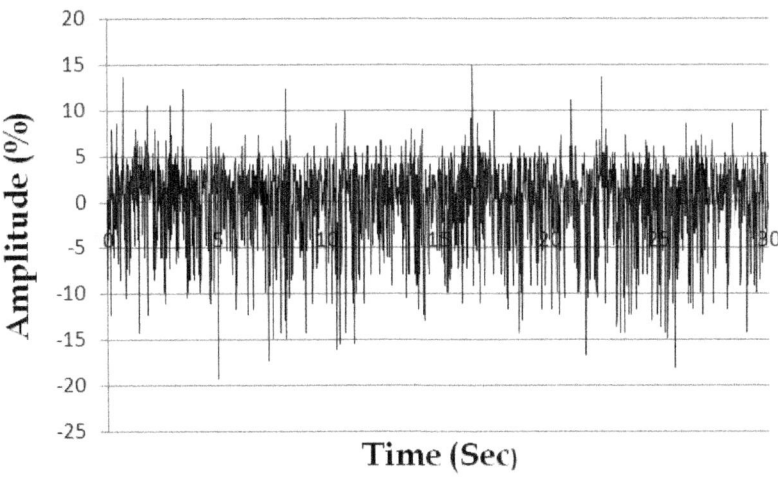

Figure 5. Effects of building vibration on the acquired signal before using damper.

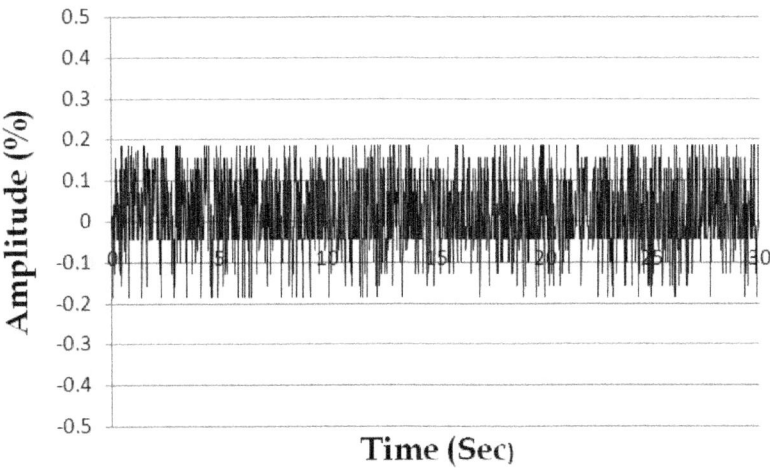

Figure 6. Effects of building vibration on the acquired signal after using damper.

In the analysis of the ground vibration, as seen, only vertical vibration of the building is considered. However, it is clearly known that buildings are exciting in three dimensions. As seen in the analysis, it was assumed that only vertical vibration had a significant effect on the data. This assumption may introduce some experimental errors into the measurement. However, a significant drop in the amplitude of the noise transmitted suggested that either the tennis balls were also eliminating some of the vibration effects coming from the other directions, or the vibration coming from the other directions did not have a significant effect on the signal. Therefore, vibration effects from other directions were not investigated further in this analysis.

In the engineering view of the problem, the ground vibration problem has a crucial effect on measurement instruments. Therefore, it would be advisable to take all of the necessary precautions in order to reduce the amount of transmitted ground vibration to the minimum level as much as possible by using damper systems which make the ratio of ω/ω_n is bigger than $\sqrt{2}$.

REFERENCES

1. D. Agres, 2007 Dynamics of Frequency Estimation in the Frequency Domain, IEEE Transactions on Instrumentation and Measurement, 56 6 (December 2007) 2111-2118 0018-9456
2. Y. Bozorgnia, V. V. Bertero, 2004 Earthquake Engineering: From Engineering Seismology to Performance-Based Engineering, CRC Press LLC, 0-20348-624-2 Raton, FL.
3. W. F. Chen, E. M. Lui, 2005 Earthquake Engineering for Structural Design, CRC Press Taylor & Francis Group, 0-84937-234-8 Raton, FL.
4. E. Chu, A. George, 1999 Inside the FFT Black Box Serial and Parallel Fast Fourier Transform Algorithms, CRC Press LLC, 0-84930-270-6 Raton, FL.
5. A. Croft, R. Davison, M. Hargreaves, 2000 Engineering Mathematics: A Foundation for Electronic, Electrical, Communications and System Engineers, Third Edition, Pearson, Prentice Hall, 0-13026-858-5 Warwickshire, UK.
6. D. Halliday, R. Resnick, J. Walker, 2007 Fundamentals of Physics Extended 7th Edition, John Wiley & Sons, Inc., 0-47147-062-7 NJ.
7. W. Kester, 2004 The Data Conversion Handbook, Elsevier, 0-75067-841-0 UK.
8. R. W. Ramirez, 1985 The FFT Fundamentals and Concepts, Tektronix, Inc. 0-13314-386-4 NJ.
9. S. W. Smith, 2006 The scientist & Engineer's Guide to Digital Signal Processing, California Technical Publishing, 0-96601-763-3 Diego, CA.
10. S. V. Vaseqhi, 2009 Advanced Digital Signal Processing and Noise Reduction: Fourth Edition, John Wiley & Sons Ltd., 978-0-47075-406-1 West Sussex, UK.
11. M. E. Yuksekkaya, W. Oxenham, 1999 Analysis of Mechanical and Electrical Noise Interfacing the Instrument During Data Acquisition:

Development of a Machine for Assessing Surface Properties of Fibers, Textile, Fiber and Film Industry Technical Conference, Atlanta, GA.

12. M. E. Yuksekkaya, 1999 A Novel Technique for Assessing the Frictional Properties of Fibers, Ph.D. Dissertation, North Carolina State University, Raleigh, NC.

13. M. E. Yuksekkaya, W. Oxenham, M. Tercan, 2008 Analysis of Mechanical and Electrical Noise Interfacing the Instrument During data Acquisition for Measurement of Surface Properties of Textile Fibers, IEEE Transactions on Instrumentation and Measurement, 57 12 (December 2008) 2885-2890 0018-9456

14. D. G. Zill, M. R. Cullen, 2006 Advanced Engineering Mathematics, Third Edition, Jones and Bartlett Publishers, 139780763745912, Sudbury, MA.

Chapter 3

THEORETICAL MODELLING AND EFFECTIVENESS STUDY OF SLOTTED STAND-OFF LAYER DAMPING TREATMENT FOR RAIL VIBRATION AND NOISE CONTROL

Caiyou Zhao[1,2] and Ping Wang[1,2]

[1]Key Laboratory of High-Speed Railway Engineering, Ministry of Education, Southwest Jiaotong University, Chengdu 610031, China

[2]School of Civil Engineering, Southwest Jiaotong University, Chengdu 610031, China

ABSTRACT

A promising means of reducing railway noise is to increase the damping of the rail, which decreases the vibration of the rail to reduce noise. To achieve this goal, a slotted stand-off layer damping treatment has been developed, and a compound track model with this treatment is developed for investigating the effectiveness of this treatment in terms of the vibration reduction. Through the dynamic analysis of the track undergoing the slotted stand-off layer damping treatment, some guidelines are proposed on the selection of materials and structure parameters for this treatment. In addition, the prototype of the optimal slotted stand-off layer damping treatment has been built and tested in the laboratory. It is found that the slotted stand-off damping treatment shows significant effects in decreasing the amplitude of the acceleration of the rail and a significant reduction of sound emission reflected as the radiation sound pressure level decreases by 8.2 and 9.4 dB at vertical excitation and lateral excitation, respectively, in the frequency range of 0–4000 Hz.

INTRODUCTION

With the development of the high-speed railway and urban rail transit, railway noise has become a growing public concern in the world today. There are plenty of noise sources associated with the railway. However, in general, railway noise is composed mainly of three types: rolling noise, traction noise, and aerodynamic noise [1]. Figure 1 presents the predominant noise sources in the total noise level depending on train speed, of which rolling noise is the greatest contributor to the overall noise level when the train is running at a speed of less than 300 km/h [2].

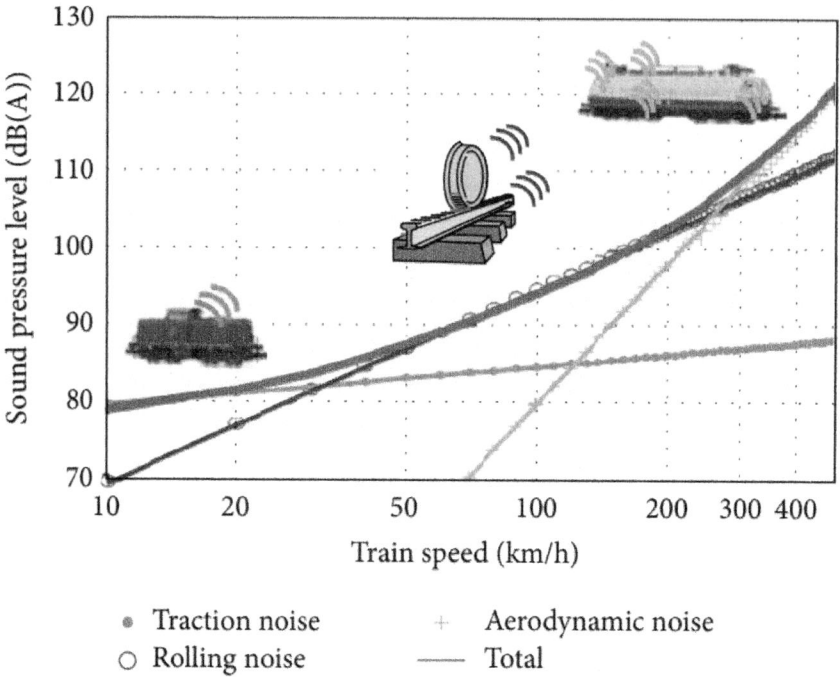

- Traction noise
- Rolling noise
- Aerodynamic noise
- Total

Figure 1: Relative strength and speed dependence of sources of railway noise [2].

Extensive work has been conducted to understand the mechanism of railway rolling noise generation [3–6]. The vibration of the wheel, rail, and sleeper caused by surface roughness of the rail and train wheel contributes to the overall spectrum of noise. Figure 2 shows, over the whole frequency range, the contribution of the three elements. It is an example of the calculations by Thompson with the TWINS software, which have been validated in extensive field measurements [7, 8]. Amongst them, the rail is a dominant source in the frequency region of 450–2000 Hz, and it generates a larger contribution to the total noise above 2000 Hz. This fact means that rail radiation is one of the significant sources of rolling noise.

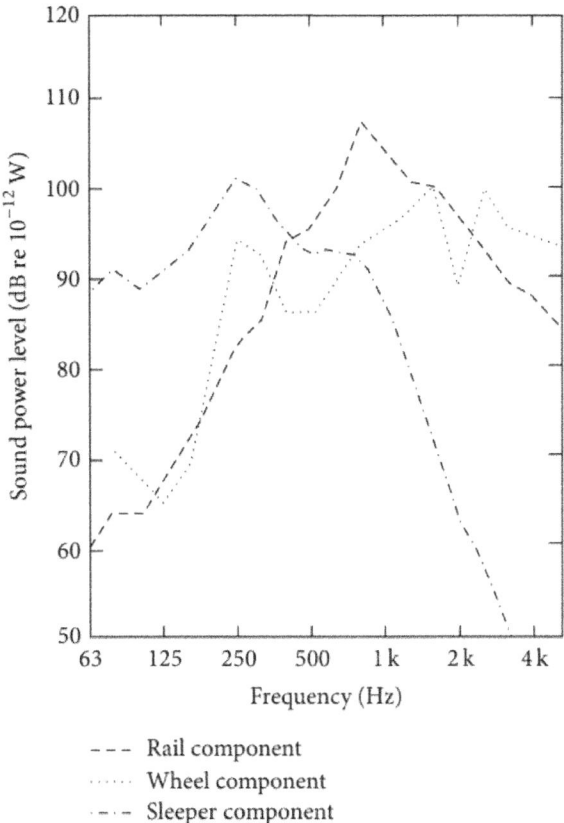

Figure 2: Noise emissions of the track [7, 8].

Numerous measurements have been taken to mitigate the acoustic emission of rails; however, the application of dampers to the rail, in the form of dynamic absorbers or superficial damping layers, is theoretically and practically the most effective. A dynamic absorber can be designed to affect only some specific frequencies so that their damping effect is concentrated within the frequency ranges of interest. Many experts have paid much attention to the application of dynamic absorbers in rails. Maes and Sol presented a double tuned rail damper, which is used to reduce the first two vertical "pined-pined" vibrations of the rail. The damper is mounted between two sleepers on the rail. Measurements at a test track of the French railway company showed that the wave decay rate of rail vibration can be effectively increased by the rail damper [9]. Ho et al. developed a new type of tuned mass damper which comprises multiple masses oscillating along the shear direction of resilient layers forming a multiple-mass-spring system and it can reduce the rolling noise of track components by 5 dB in field tests [10]. Ltourneaux et al. studied another rail damper, each unit of which was shorter but wider than Thompson's damper. A reduction of 3 dB in overall noise was found in tests on a running line with pads of medium stiffness, approximately 350 kN/mm [11]. Asmussen et al. produced a rail damper that consisted of steel masses and elastomer-based materials. The elastomer was used to absorb energy through internal friction as the steel elements vibrate in response to vibrations in the rail. On a test track, a noise reduction up to 4 dB, due to the increased damping of the rail, was measured [12].

While dynamic absorbers affect only some specific frequencies, superficial damping layers can be designed to act on the entire frequency range, depending on the viscoelastic property of the damping material. Similarly, many studies have been conducted in the past to investigate noise abatement characteristics of superficial damping layers and their applications on rails. Wei et al. developed a compound damping board, a complex of high damping material and restrained board. An experimental study on the dynamic behavior of rail attached to the compound damping board has been presented. The results of field tests prove that the compound damping board can reduce the lateral vibration level of rails by 6-7 dB [13]. Li et al. built a constrained damping layer that was made of a damping layer (2 cm thick) and an aluminium constrained layer (3 cm thick). Research results showed that these damped rails had a good effect on vibration reduction in a wide frequency range of 0–5000 Hz [14]. Xue Jun offered a labyrinth constrained damping layer that consisted of three parts: a connection layer, a damping layer, and a constrained layer. The field experiment results showed that the labyrinth constrained damping layer can reduce rail noise by up to 7 dB [15]. In general, these studies have focused on the application of traditional superficial free-damping layers or superficial constrained damping layers to dissipate rail vibration energy, reducing the acoustic emission of rails. However, little effort has been devoted to the application of more efficient damping treatment measurement in rails.

The research presented herein constitutes an exploratory study of the slotted stand-off layer damping treatment for rail vibration and noise control using both theoretical analysis and laboratory tests. First, the working principle of the slotted stand-off layer damping treatment is discussed. Then, a compound track model with this treatment is developed for investigating the effectiveness of this treatment in terms of vibration reduction. Additionally, some guidelines are provided on the selection of materials and structure parameters by the model for this treatment. Finally, the optimal slotted stand-off layer damping treatment is used to make a full scale sample. The vibration and acoustic reduction effect is measured in the laboratory.

WORKING PRINCIPLE OF THE SLOTTED STAND-OFF LAYER DAMPING TREATMENT

Applying a surface damping treatment to a stiff beam, plate, shell, and other thin-walled structures can reduce the amplitude of a vibration [16]. When the damping treatment is applied as a single-layer coating, sometimes known as a free-layer damping, energy is dissipated as a result of the extension and compression of the damping material while the base structure bends during vibration. A significant increase in energy dissipation can be achieved by attaching a stiff layer (or constraining layer) on top of the viscoelastic layer, as shown in Figure 3. This dissipation occurs because large shear strains are generated in the damping material when the system bends during vibration. Adding a stand-off layer to the constraining damping treatment between the base layer and the damping layer could further increase the shear angle of the damping layer. This damping treatment has been termed passive stand-off layer damping treatment. However, for the effective operation of passive stand-off layer damping treatment, the stand-off layer must have high shear stiffness and, at the same time, must not significantly affect the bending stiffness of the composite structure. Based on this principle, in one variation of passive stand-off layer damping treatment, slots are introduced into the stand-off layer. Slotted stand-off layer damping treatment, as shown in Figure 4, has been thought to have an important advantage over continuous passive stand-off layer damping treatment [17].

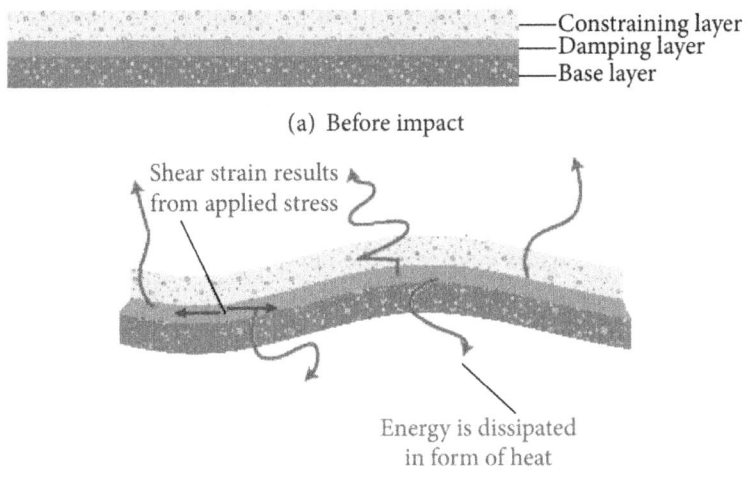

Figure 3: Working principle of a constraining damping treatment system.

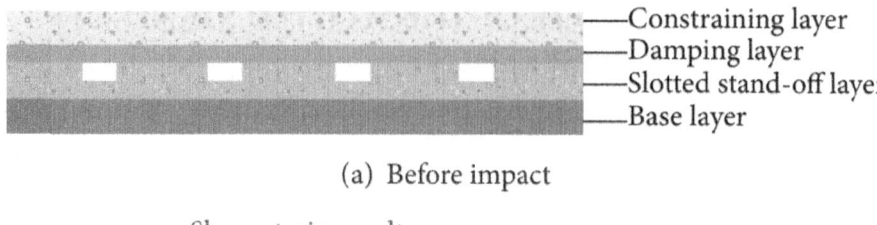

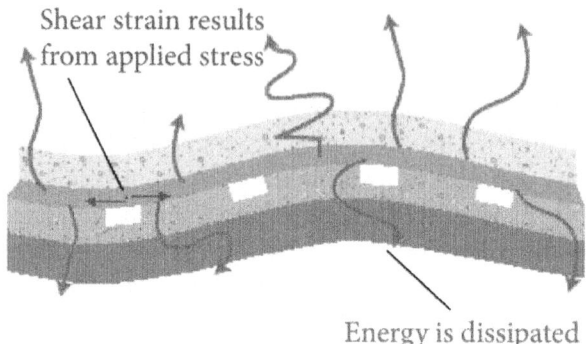

(a) Before impact

(b) Being deformed during vibration

Figure 4: Working principle of a slotted stand-off layer damping treatment.

In general, the stand-off layer has no internal damping properties. Therefore, the damping ratio h is used to assess the vibration reduction effect of the stand-off layer damping treatment as the traditional constraining damping treatment. The damping ratio h can be defined, according to the "strain energy method," as the ratio between the energy dissipated by the damping layer in each strain cycle and that stored in the system:

$$\eta = \frac{D_0}{2\pi W_0}, \quad (1)$$

where D_0 is the dissipated energy in the damping layer and W_0 is the stored energy of the system in a strain cycle.

As for D_0,

$$D_0 = \pi \cdot G''(\omega) \cdot V \cdot \gamma^2, \quad (2)$$

where g is the shear strains of the damping layer, $G''(\omega)$ is the loss modulus of the damping material, and is the volume of the damping layer. The addition of a slotted stand-off layer only increases shear strains, thereby reducing the vibration of the base layer more effectively.

OPTIMISATION ANALYSIS OF SLOTTED STAND-OFF LAYER DAMPING TREATMENT

Some studies have been conducted in the past to investigate the application of stand-off layer damping treatment in industrial fields and the corresponding calculation of their damping ratios. Generally, the theoretical solution for the loss factor calculation is unfortunately available only for simple geometry structures. More recently, an alternative approach, based on numerical finite element calculations, allowed the consideration of a different component shape, as in the case of a rail. Furthermore, in the slotted stand-off layer damping treatment system, the bonding layers used to connect these layers are found to have a measurable and significant effect on the response of the structure. Therefore, the finite element model presented here includes an epoxy layer between the base layer and the slotted stand-off layer and a contact cement layer between the slotted stand-off layer and the damping layer. The six-layered model incorporating a primer layer and an epoxy layer, as shown in Figure 5, proposed by Yellin et al. [18], is adopted in this study to model slotted stand-off layer damping treatment.

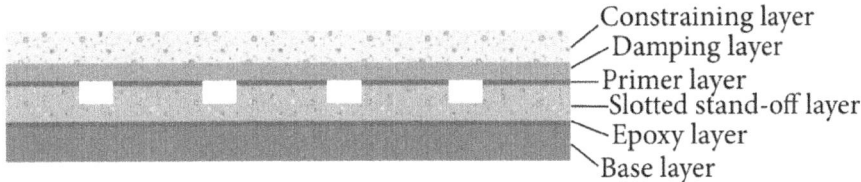

Figure 5: Schematic diagram of slotted stand-off layer damping treatment.

Numerical Procedure for the Damping Ratio Calculation

In the present work, an approach based on a steady-state harmonic analysis in the frequency range of interest has been adopted. This approach directly integrates the equations of motion of the system. Thus, all of the vibration modes involved at any excitation frequency will be taken into consideration. This approach, therefore, simulates the actual behavior of the component. To this end, a postprocessing procedure for the ANSYS commercial code has been developed and can be generally applied to any component shape and material. In particular, the dependence of the elastic and viscous properties of the damping layer material on frequency and temperature will be taken into account to define these properties directly as supplied by the producer. The results calculated in this way are very flexible.

The specific procedure calculates the damping ratio h in the discretized form.

(a) The dissipated energy D_0 mainly derives from the shear deformations of the damping layer:

$$D_0 = \sum_i (D_0)_i = \sum_{i=1} 2\pi W''_{shear_i}, \qquad (3)$$

whereby

$$W''_{shear_i} = \iiint \frac{G''(\omega)}{2} \sum_{j=1}^{2} \gamma_{0j}^2 dx\, dy\, dz, \qquad (4)$$

so,

$$D_0 = \pi \cdot G''(\omega) \sum_i \left[V_i \sum_{j=1}^{2} (\gamma_{0j}^2)_i \right]. \qquad (5)$$

(b) The stored energy W_0 mainly derives from two parts, namely, the stored energy from the damping layer generated by shear deformations and the stored energy from the other five layers generated by elastic deformations:

$$W_0 = \sum_i (W_0)_i = \sum_i (W_S + W_V + W'_{shear}), \qquad (6)$$

where WS, WV, and $W'shear$ denote the elastic deformation energy.

For the ith element ($i = 1 \ldots n$),

$$W_i = \frac{1}{2}(\sigma_1 \varepsilon_1 + \sigma_2 \varepsilon_2 + \sigma_3 \varepsilon_3)$$

$$= \frac{1}{2E}\left[\sigma_1^2 + \sigma_2^2 + \sigma_3^2 - 2v(\sigma_1 \varepsilon_1 + \sigma_2 \varepsilon_2 + \sigma_3 \varepsilon_3)\right],$$

$$W_{Vi} = \frac{1}{2} \cdot \frac{\sigma_1 + \sigma_2 + \sigma_3}{3}(\varepsilon_1 + \varepsilon_2 + \varepsilon_3)$$

$$= \frac{1 - 2v}{6E}(\sigma_1 + \sigma_2 + \sigma_3)^2$$

$$= \frac{3(1 - 2v)}{2E}\sigma_{oct}^2,$$

$$W_{Si} = W_i - W_{Vi}$$

$$= \frac{1}{2E}\left[\sigma_1^2 + \sigma_2^2 + \sigma_3^2 - 2v(\sigma_1\sigma_2 + \sigma_2\sigma_3 + \sigma_1\sigma_3)\right.$$

$$\left. -\frac{1-2v}{3}(\sigma_1 + \sigma_2 + \sigma_3)^2\right]$$

$$= \frac{1}{12G}\left[(\sigma_1 - \sigma_2)^2 + (\sigma_2 - \sigma_3)^2 + (\sigma_1 - \sigma_3)^2\right]$$

$$= \frac{1}{6G}\sigma_{Mises}^2$$

$$= \frac{1+v}{3E}\sigma_{Mises}^2,$$

$$W'_{shear_i} = \frac{G'(\omega)}{2}\sum_{j=1}^{2}\gamma_{0j}^2. \tag{7}$$

Substituting (7) into (6) gives

$$W_0 = \sum_i V_i \left[\frac{1+v}{3E}\sigma_{Mises_i}^2 + \frac{3(1-2v)}{2E}\sigma_{oct_i}^2\right.$$

$$\left. + \frac{G'(\omega)}{2}\sum_{j=1}^{2}(\gamma_{0j}^2)_i\right], \tag{8}$$

where V_i is the volume of the ith element, g_{0j} is the jth shear deformation of the ith element, s_{Misesi} is the von Mises stress of the ith element, s_{Octi} is the octahedral stress of the ith element, G' is the storage shear modulus, G" is the loss shear modulus, E is Young's modulus of elasticity, and n is the Poisson ratio.

The damping ratio h can be obtained by substituting (5) and (8) into (1).

Choice of the Material and Configuration of the Slotted Stand-Off Layer Damping Treatment

Several commercial damping materials were examined in this project. One of these, VER-IPN, was chosen because of its high loss-factor values in the temperature and frequency ranges of interest, as depicted in Figure6. The slotted stand-off layer was made from Dyad606 with high shear strength. Its density and Young's modulus of elasticity were 1200 kg/m³ and 500 MPa, respectively, as tested. Although the epoxy layer and primer layer were relatively thin, they should be as stiff as possible; therefore, a strong adhesive that has good synthetic properties was developed. The adhesive had a tensile modulus of approximately 1.65 GPa. The material used for the constraining layer can be either Al-2A14, with an elastic modulus of E=73.1 GPa, or steel-3Cr13Mo, with an elastic modulus of E=206 GPa.

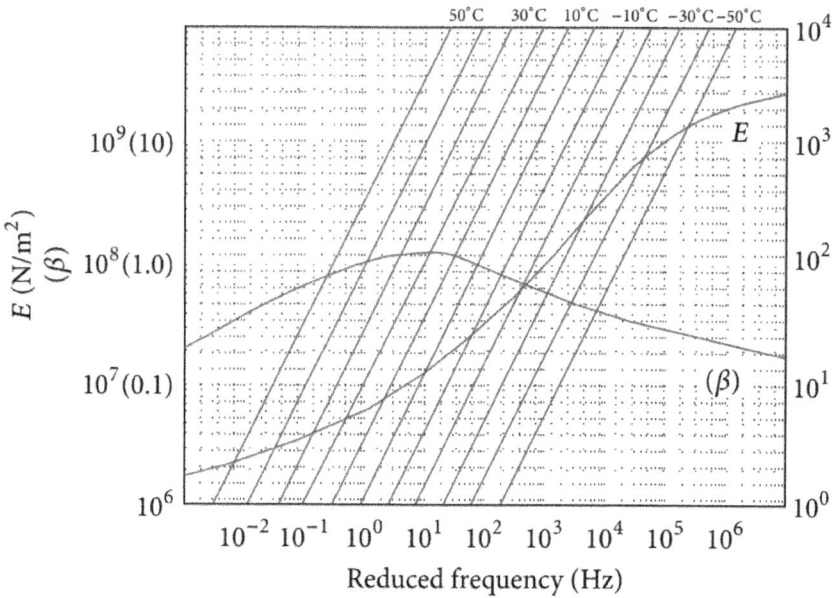

Figure 6: Damping VER-IPN property specification.

However, it is not easy to choose the best arrangement for slotted stand-off damping treatment. The rail waist and the rail foot are the weakest zones of the rail; therefore, such parts appear to be the most feasible in terms of application and the most promising in terms of effectiveness. However, this solution shows difficulties in shaping the damping layer to match the rail profile. A 60 kg/m rail, which is the most widely used in China, is studied in this paper. The basic idea is to apply a slotted stand-off damping treatment to specified zones of a widely used rail, without additional machining operations on the rail itself. Several solutions were initially proposed and examined, taking the following design requirements into consideration: damping efficiency, safety, and reliability under working conditions (in particular mechanical resistance, weathering resistance, moisture and solvent resistance, and thermal and fire resistance), ease of manufacturing and application, weight, and cost. Finally, the slotted stand-off layer damping treatment is arranged according to the special structure of the rail, as depicted in Figure 7, consisting of four major parts: an inverted T-shaped constraining layer of uniform thickness (3), connected to the rail with a slotted stand-off layer with uniform thickness (1) and sandwiched between the slotted stand-off layer and the constraining layer with a slot with a thickness-variable damping layer(2); the rail(4)is surrounded by stand-off layer. The strong adhesive is applied to each layer.

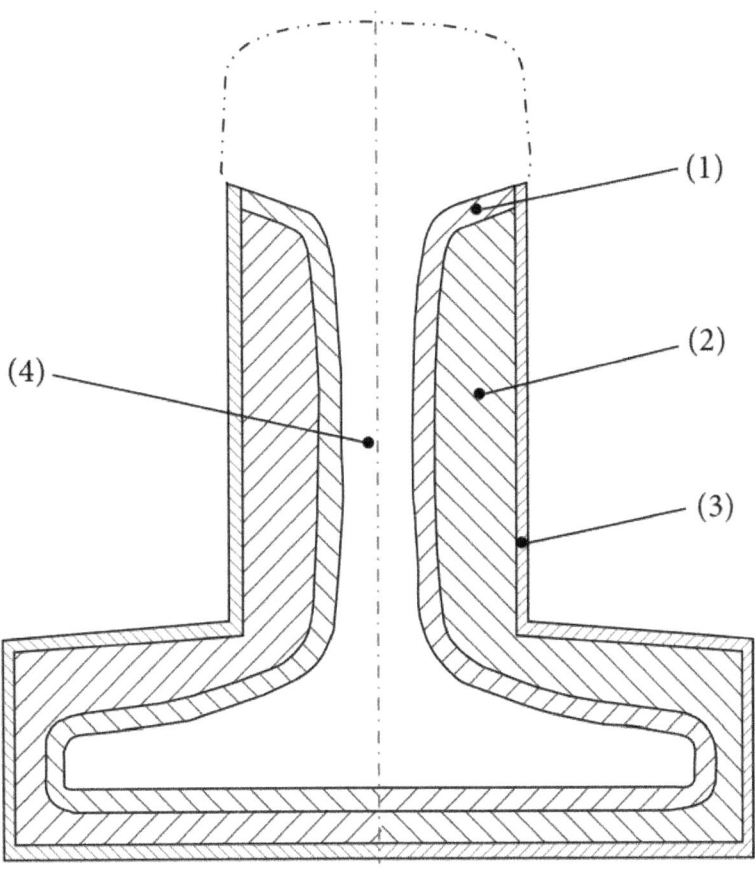

(a) Without slot

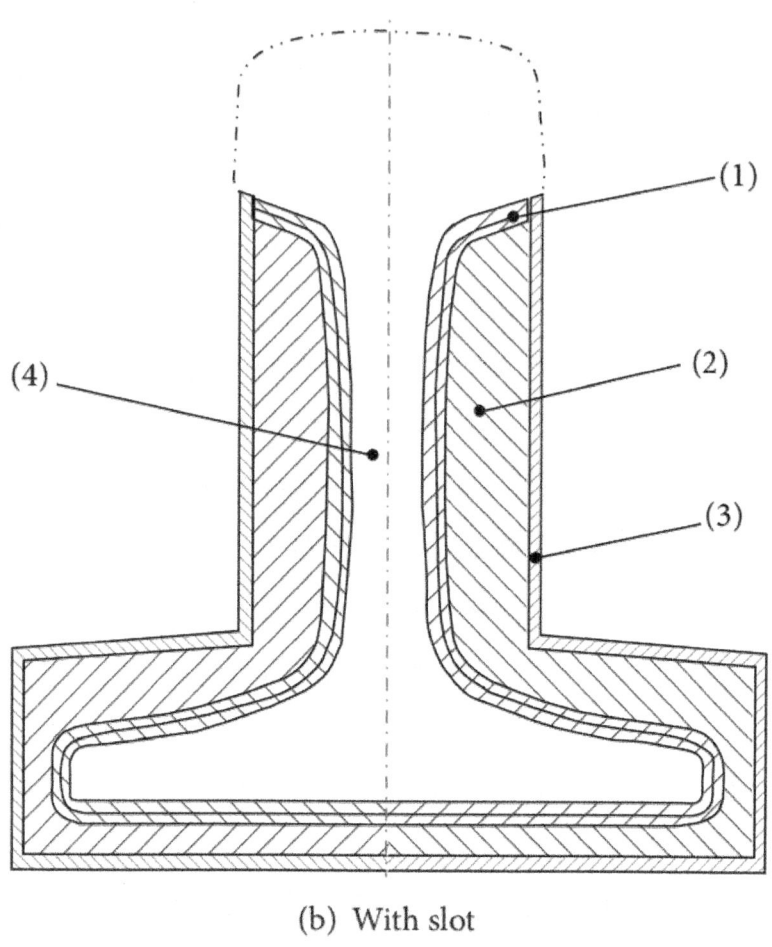

(b) With slot

Figure 7: Cross section of the rail with the slotted stand-off layer damping treatment.

The numerical procedure described above was used to examine the effectiveness of different configurations. The modifiable structural size of the slotted stand-off layer damping treatment was defined, as shown in Figure 8. In particular, the calculations were carried out for the following seven cases (also shown in Table 1), where h_{s1} is the zone thickness of the slotted stand-off layer without slots, h_{s2} is the slot thickness of the slotted stand-off layer, t_{s1} is the slot width, t_{s2} is the zone width of the slotted stand-off layer without slots, h_c is the thickness of the constraining layer, $t_{c1} \sim t_{c3}$ are the three important structural sizes of the constraining layer, and α is the slope of the constraining layer.

Table 1: Tests of the material and configuration

Case	h_{s1}/mm	t_{c1}/mm	α	t_{c2}/mm	t_{c3}/mm	h_c/mm	h_{s2}/mm	t_{s2}/mm	t_{s1}/mm	Damping layer material	Constraining layer material
1	8	100	4%	55	45	3	2	30	20	VER-IPN	Al-2A14
2	6	100	4%	55	45	3	2	30	20	VER-IPN	Al-2A14
3	6	100	4%	55	45	3	3	30	20	VER-IPN	Al-2A14
4	6	100	4%	55	45	3	3	20	20	VER-IPN	Al-2A14
5	6	100	4%	55	45	3	3	20	30	VER-IPN	Al-2A14
6	6	100	4%	55	45	2	3	20	30	VER-IPN	Al-2A14
7	6	100	4%	55	45	2	3	20	30	VER-IPN	Steel-3Cr-13Mo

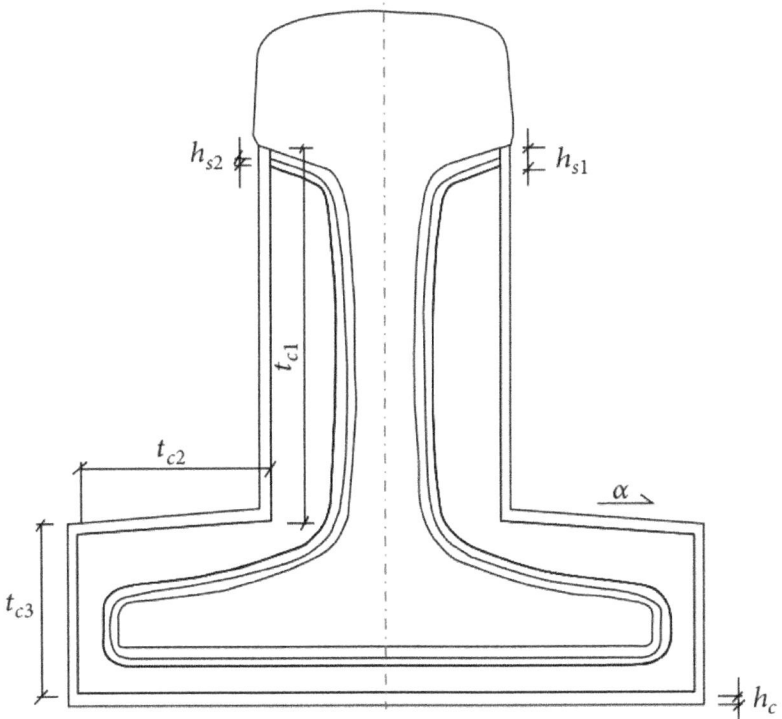

(a) Cross section of the rail with damping treatment

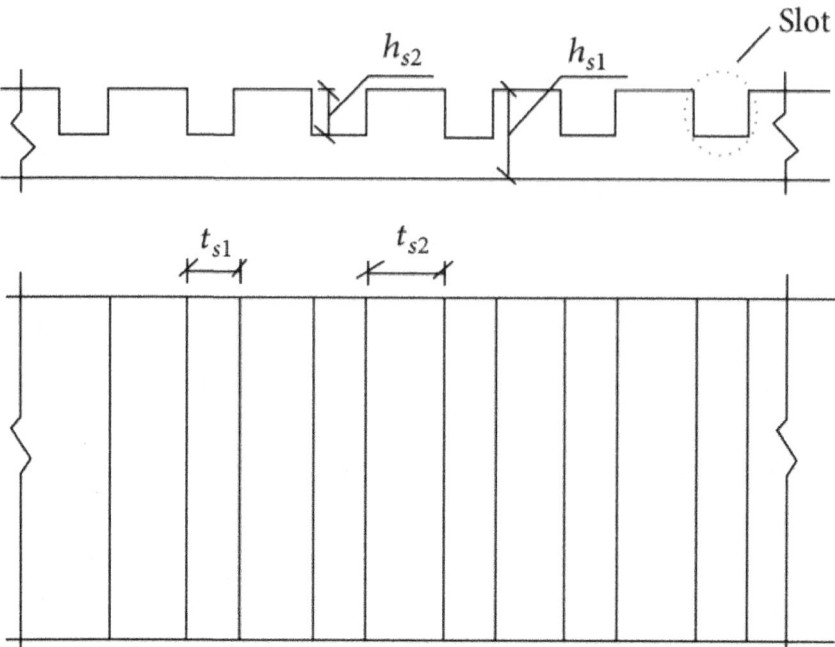

(b) Detailed characteristic of slotted stand-off layer

Figure 8: Parameter definitions of the slotted stand-off layer damping treatment.

Previous research has shown that the length of a rail mode with twelve spans can yield the whole resonance and antiresonance characteristics of a rail in the frequency range of 0~5000 Hz [19]. Thus, the length of the rail mode with twelve spans (with a total length of 7.2 m) is applied in the present work. The rails are fastened by the WJ-7 flexible fastening systems every 0.6 m, and the fastening systems are fixed on the prestressed reinforced concrete sleepers III. Moreover, the operating temperature of the damping material is assumed to be 30°C. To represent realistic working conditions, two forcing actions, one twice as much as the other, are applied simultaneously on the railhead surface in the vertical and lateral directions, as shown in Figure 9; these actions simulate the typical main loads transmitted from the wheel to the rail (11).

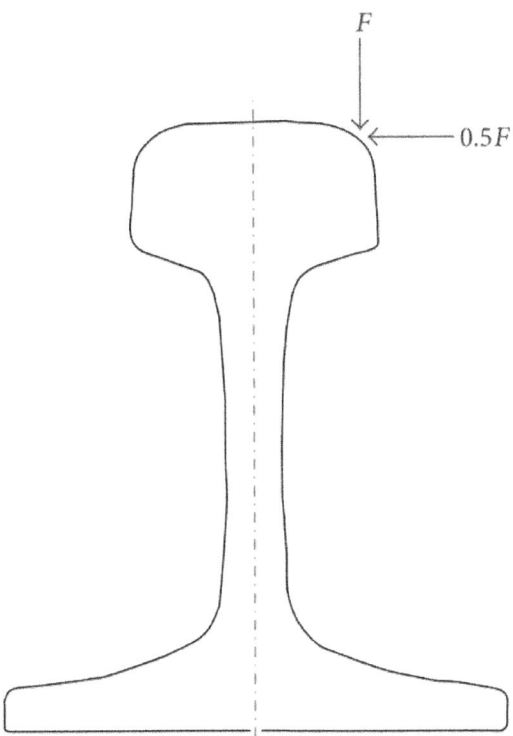

Figure 9: Schematic plot of the forcing actions.

The force F is calculated as follows:

$$F = (1 + \alpha_1) \cdot (1 + \alpha_2) \cdot P, \tag{9}$$

where $\alpha_1 = 0.6\dfrac{v}{100}$, $\alpha_2 = 0$, (10)

when v is less than 120 km/h,

$$\alpha_1 = 0.72, \quad \alpha_2 = 0.3\dfrac{\Delta v}{100}, \quad \Delta v = v - 120, \tag{11}$$

when v is larger than 120 km/h,
where v is the vehicle speed and P is the axle load of the vehicle.

The axle load P of 15 T (CRH3 high-speed train) is studied in this paper, and the speed of the train is 300 km/h. Therefore, the force F is 397,270 N.

RESULTS

Thirteen representative frequencies were chosen from the modal analysis of the track model described above. The damping ratios calculated from the seven different cases are shown in Table 2. The following can be deduced from Table 2.

- The damping ratio η of the rail with the slotted standoff layer damping treatment is mostly greater than 0.017. However, the η of the standard rail is only 0.0001~0.0006. This result means that the damping ratio η of the rail with the slotted stand-off layer damping treatment far outweighs the η of the standard rail.

- The zone thickness of the slotted stand-off layer without slots imposes different influences on specific eigenmodes, as noted by comparing the results of Cases 1 and 2. The thickness of Case 1 is more effective for the first, third, fifth, sixth, eighth, and thirteen eigenmodes, whereas that of Case 2 is more effective for the other eigenmodes. Furthermore, in view of the great rigidity of the material of the slotted stand-off layer, as mentioned, it is more reasonable to apply the thickness from Case 2.

- The structural dimensions of the slot exert a strong influence on the damping ratios, as highlighted by comparing the results of Cases 2 and 3 and Cases 4 and 5: the thicker and wider the slot, the greater the damping ratios in the frequency range of greatest interest. The influence of the zone width of the slotted stand-off layer without the slot is contrary to the structural dimensions of the slot: the wider the zone of the slotted stand-off layer without slots, the smaller the total width of the slot.

- The thickness of the constraining layer has some effect on the damping ratio η of the rail with slotted standoff layer damping treatment, as shown by comparing the results of Cases 5 and 6. Case 5 has damping ratios slightly larger than that of Case 6 in addition to the fourth, sixth, seventh, and ninth eigenmodes. However, a 3 mm thick aluminium plate or steel plate is more difficult to be machined and shaped than 2 mm plates. Therefore, it is better to adopt the 2 mm thick constraining layer. The difference between steel and aluminium constraining layers (Cases 6 and 7) is significant, and the damping ratio of the former is obviously greater than that of the latter.

- In sum, to obtain as great of a noise reduction as possible in the frequency range of interest and to manufacture the system as easily as possible, Case 7 is the best choice.

Table 2: Damping ratio η of the seven cases

Frequency	Case 1	Case 2	Case 3	Case 4	Case 5	Case 6	Case 7
100	0.018	0.017	0.019	0.021	0.025	0.024	0.028
200	0.022	0.023	0.024	0.026	0.031	0.031	0.035
600	0.033	0.031	0.034	0.035	0.044	0.042	0.044
1000	0.041	0.042	0.044	0.048	0.059	0.061	0.065
1300	0.132	0.128	0.119	0.129	0.163	0.156	0.187
1500	0.209	0.231	0.221	0.248	0.287	0.293	0.358
2100	0.164	0.163	0.171	0.187	0.212	0.218	0.269
2500	0.132	0.128	0.162	0.158	0.176	0.175	0.218
2700	0.121	0.130	0.152	0.175	0.185	0.194	0.191
3100	0.108	0.121	0.132	0.154	0.146	0.144	0.132
3400	0.290	0.321	0.414	0.374	0.392	0.389	0.432
3700	0.415	0.436	0.356	0.498	0.564	0.544	0.614
4000	0.479	0.387	0.412	0.453	0.534	0.512	0.587

LABORATORY MEASUREMENTS OF VIBRATION AND NOISE

Laboratory measurements of accelerance and noise were carried out on the Chinese 60 kg/m standard rail with the slotted stand-off layer damping treatment according to Case 7 to test its actual effectiveness compared with that of the standard rail. For the sake of economy and convenience, the length of the standard rail with the slotted stand-off layer damping treatment was 0.6 m, as shown in Figure 10.

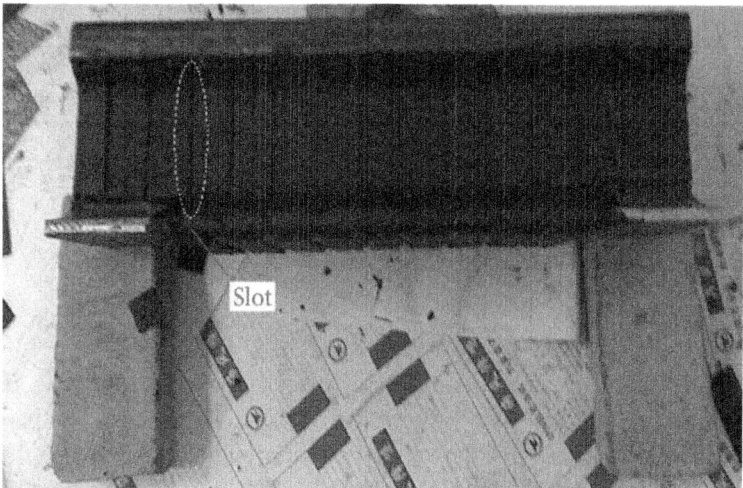

Figure 10: Schematic plot of rail with the slotted stand-off layer damping treatment.

The characteristics of this slotted stand-off layer damping treatment are as follows.
- It has a total mass of 12 kg. Because the sleeper spacing is assumed to be 0.6 m, the mass per meter of rail length is 20 kg.
- It is designed to be fitted on the rail without any modifications made to the rail, to the sleepers, or to the supporting ballast layer.
- The adhesive strength of the efficient noise-reduction glue is proven to be effective after 2 million cycles of fatigue testing.

- The electrical isolation is maintained.
- The radiation of sound by the slotted stand-off layer damping treatment itself is negligible.

Laboratory Measurements and Analysis of Accelerance

The model used for laboratory measurements and analyses of accelerance of the rail with the slotted stand-off layer damping treatment and standard rail is presented below. The rail grid model was constructed with prestressed reinforced concrete sleepers III, on which there were fastened rails with the slotted stand-off layer damping treatment, or 60 kg/m standard rails using WJ-7 flexible fastenings. For the testing of the acceleration admittance of the samples, the method of measuring the response to mechanical shock was used in accordance with ISO 7626-5 [20]. Mechanical shock was stimulated using an impulse hammer in the vertical and lateral direction on the railhead. A part of this hammer is a force detector. The response was measured using accelerometers at different points of the rail structure, on the rail head, rail waist, and rail foot of the midpoint of the rail-specimen, as shown in Figure 11.

Figure 11: Treatment configurations examined.

The measuring system consisted of a DHDAS-5920 PULSE modulator analyzer for recording the vibration parameters together with three LC0102 acceleration detectors and a LC1303 shock stimulation hammer. The accelerometers were fastened to the measured construction by means of epoxy resin adhesives.

A comparison of the vertical and lateral accelerance of the rail head between the standard rail and the rail with the slotted stand-off damping treatment is depicted in Figures 12(a) and 12(b), respectively. In Figure 12(a), it is possible to observe a general decrease in the vertical accelerance amplitude of the rail head of the rail with the slotted stand-off damping treatment amplitude in the frequency range of 0–4000 Hz, relative to the case without the slotted stand-off damping treatment, as the maximum acceleration reading decreased from $0.29\,\text{m·s}^{-2}/\text{N}$ to $0.078\,\text{m·s}^{-2}/\text{N}$. From Figure 12(b), the lateral accelerance amplitude of the rail with the slotted stand-off damping treatment is approximately $0.0015\,\text{m·s}^{-2}/\text{N}$ in the low frequency range of 0–200 Hz and the middle frequency range of 200–1000 Hz. This result means that the lateral vibration of the rail head in a frequency range of 0–2000 Hz is almost completely damped out by the slotted stand-off damping treatment. Additionally, it can be observed from the figure that compared to the standard rail, the lateral accelerance amplitude of the rail with the slotted stand-off damping treatment decreased from $0.05\,\text{m·s}^{-2}/\text{N}$ to $0.019\,\text{m·s}^{-2}/\text{N}$ in the high frequency range of 1000–4000 Hz. Thus, it was possible to obtain significant reductions in the vertical and lateral vibration levels of rail head by taking the slotted stand-off damping treatment approach in the frequency range of 0–4000 Hz.

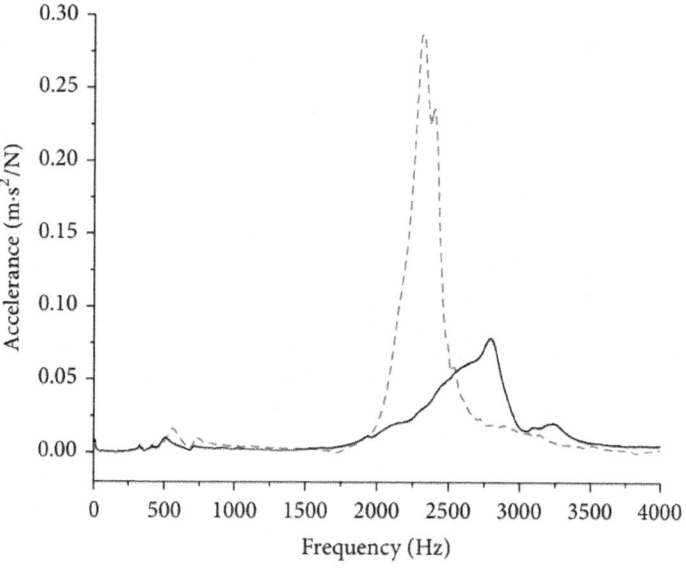

--- Standard rail
—— Rail with the slotted stand-off damping treatment

(a) Vertical accelerance of the rail head

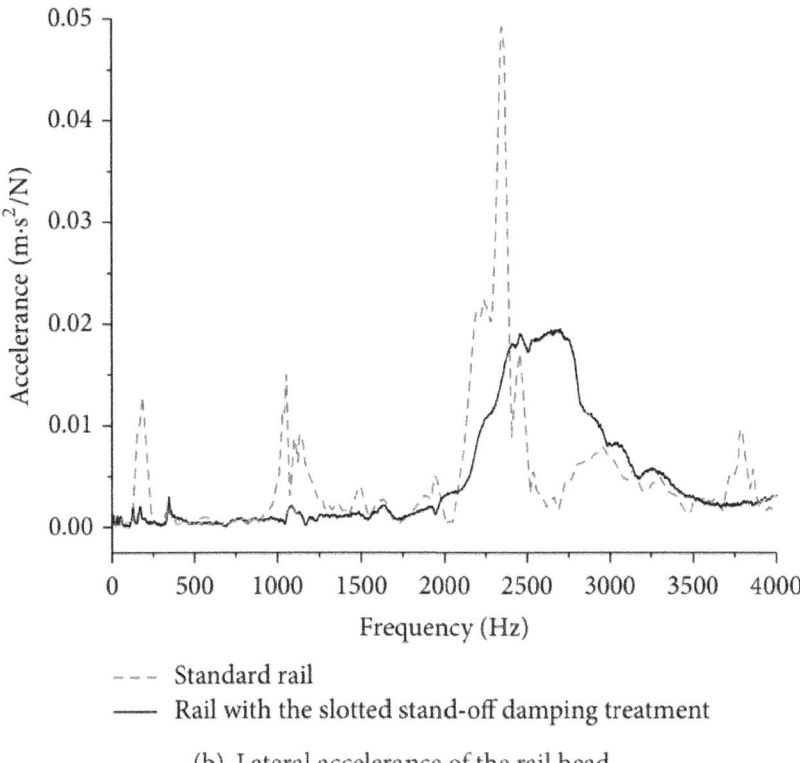

(b) Lateral accelerance of the rail head

Figure 12: Accelerance of the rail head.

Similar conclusions are observed in Figures 13 and 14, which present the analysis of the comparison of the accelerance of the rail waist and the rail foot between the standard rail and the rail with the slotted stand-off damping treatment. It can be observed from these figures that the performance of the slotted stand-off damping treatment placed at the surface of the rail for attenuating the vibrations of the rail waist and rail foot is quite effective in the frequency range of 0–4000 Hz.

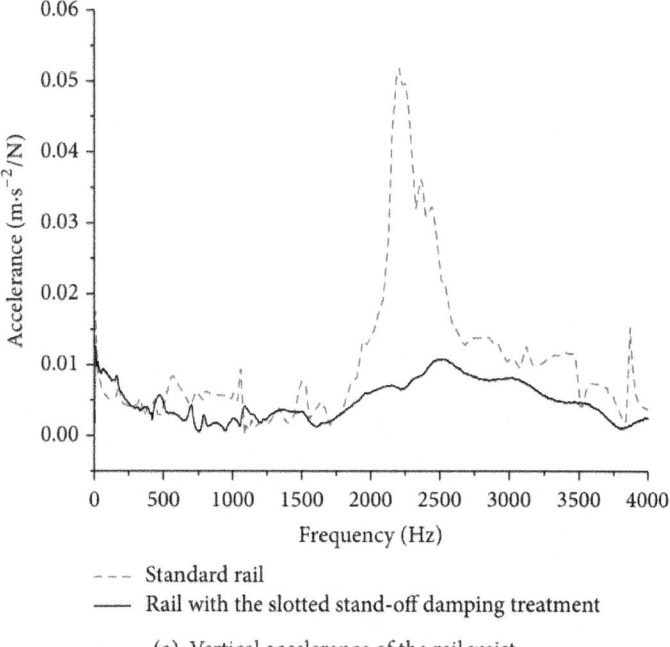

(a) Vertical accelerance of the rail waist

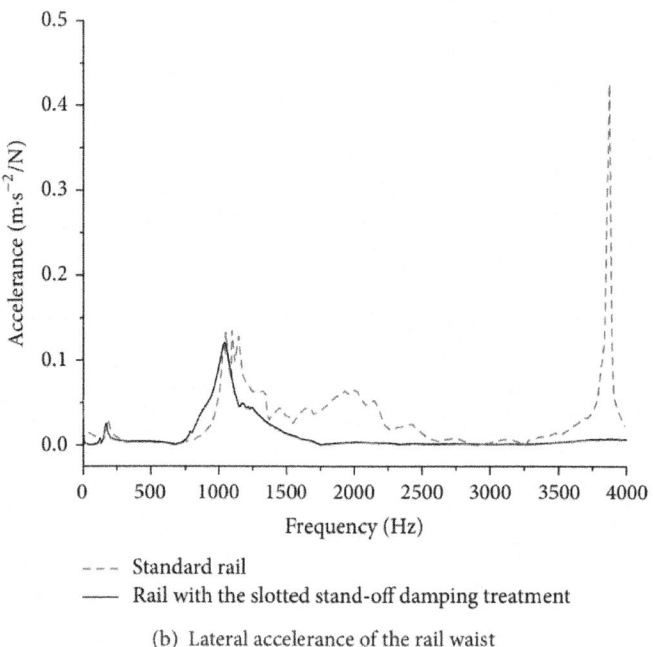

(b) Lateral accelerance of the rail waist

Figure 13: Accelerance of the rail waist.

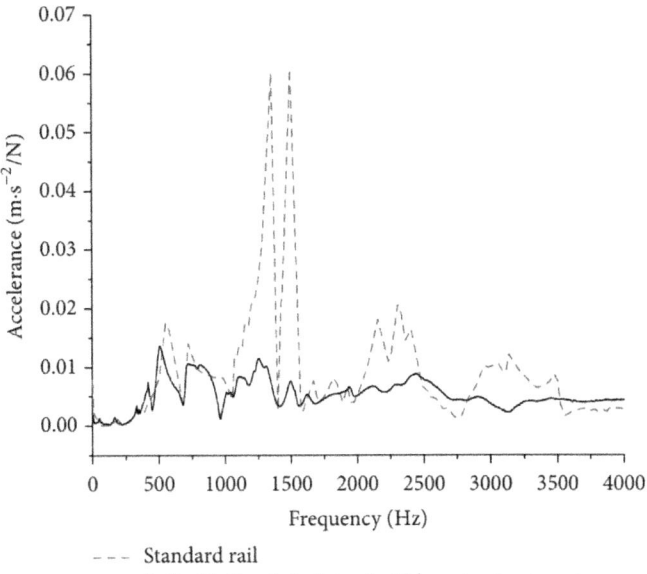

(a) Vertical acceleration of the rail foot

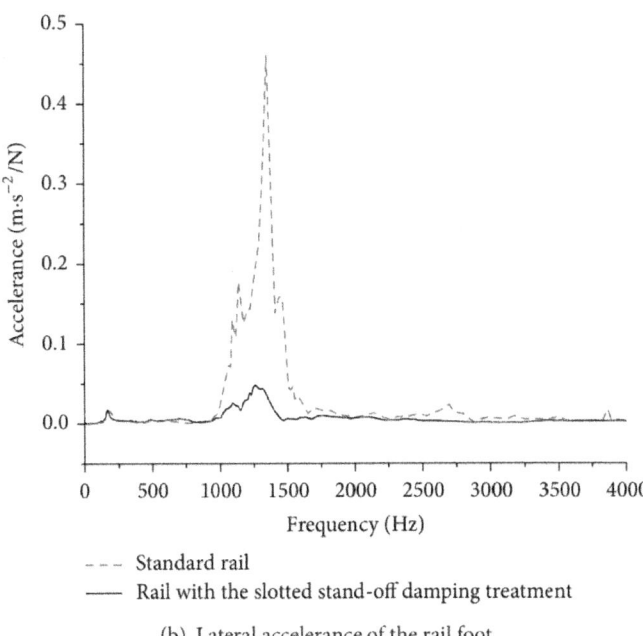

(b) Lateral acceleration of the rail foot

Figure 14: Accelerance of the rail foot.

Based on Figures 12, 13, and 14, it is possible to conclude that the slotted stand-off damping treatment decreases the amplitude of the accelerance of the rail in the frequency range of 0–4000 Hz, especially in the middle frequency range of 200–1000 Hz and the high frequency range of 1000–4000 Hz.

Laboratory Measurements and Analysis of Noise

The measurement procedure was conducted according to ISO 3744 for the source sound power emission calculation in free-field conditions by sound pressure measurements [21]. In this case, the sound power, which depends on the exciting force, was normalized by this force. The acoustic measurements were carried out in a semianechoic room according to the engineering method of ISO 3744 in the frequency range of 0–4000 Hz.

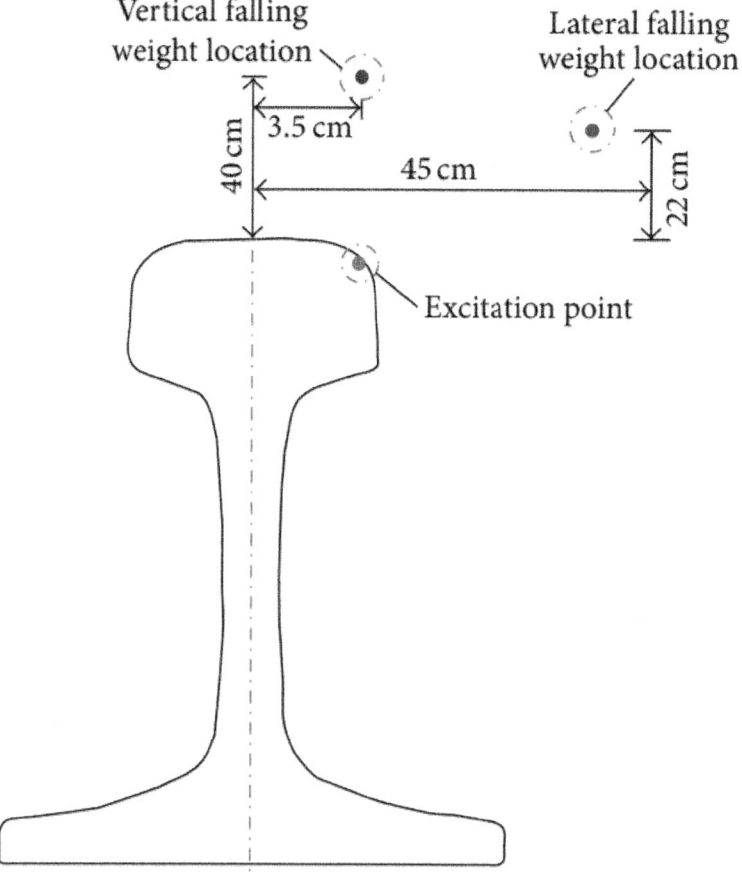

Figure 15: Location of the falling weight.

The model and its installation are the same as those applied in Section 5.1. The exciting force is an impulse caused by the falling weight (250 g). The vertical falling location is fixed over the midpoint of the rail-specimen along its length, 40 cm away from the rail head in the vertical direction and 3.5 cm away from the rail centreline in the lateral direction. Meanwhile, the lateral falling location is fixed over the midpoint of the rail-specimen along its length, 22 cm away from the rail head in the vertical direction and 45 cm away from the rail centreline in the lateral direction, as shown in Figure 15. The measurement was made using a microphone mounted on the midpoint of the rail-specimen along its length and 50 cm away from the rail centreline in the lateral direction, as shown in Figures 16 and 17.

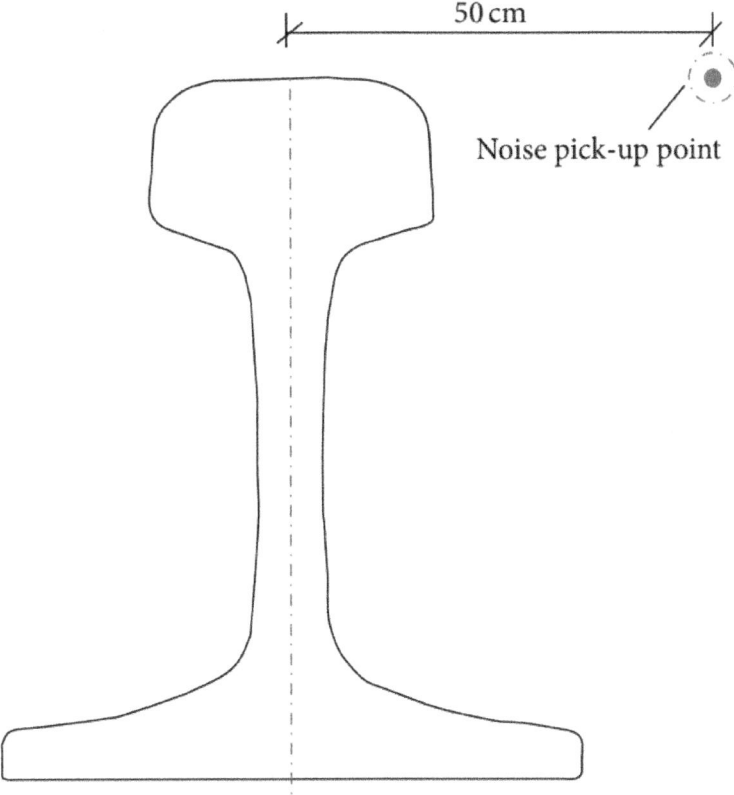

Figure 16: Location of the noise pick-up point.

Figure 17: Laboratory vibroacoustic analysis.

The laboratory results for the rail with the slotted stand-off damping treatment showed a significant reduction in sound emission compared to the standard rail under the vertical falling weight excitation, as shown in Figure 18; the radiation sound pressure level decreases by 8.2 dB. It can also be observed from Figure 19 that the rail with the slotted stand-off damping treatment produced less noise compared to the standard rail under the lateral falling weight excitation, with differences of up to 9.4 dB. Furthermore, within the frequency range of 200–4000 Hz, the noise level reduction is greater than that within the frequency range of 0–200 Hz under both vertical and lateral falling weight excitation.

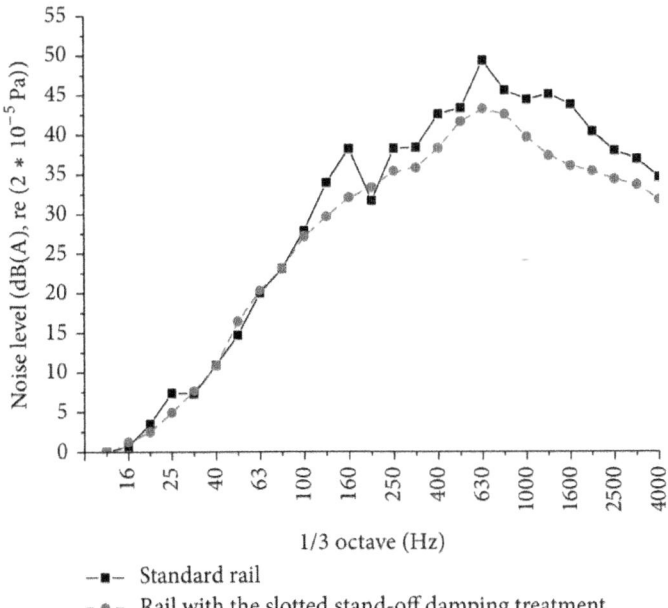

Figure 18: One-third octave curve in vertical excitation.

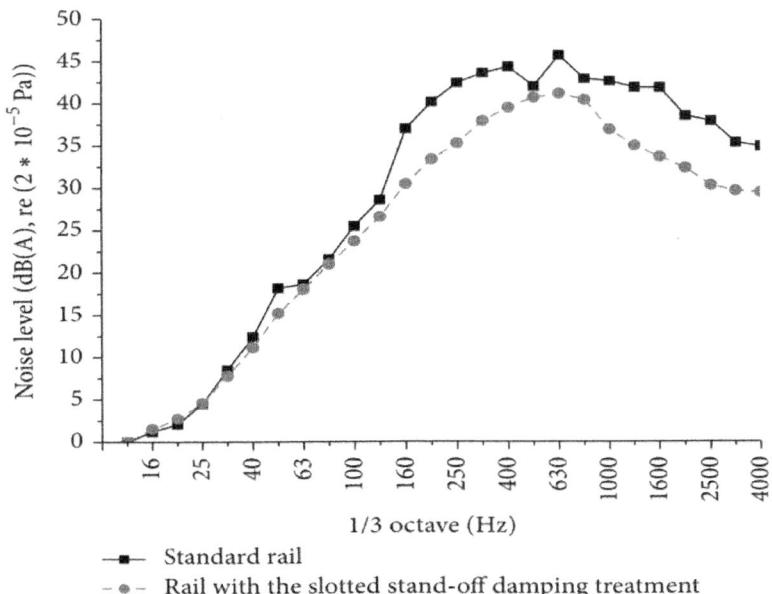

Figure 19: One-third octave curve in lateral excitation.

CONCLUSIONS

The aim of this work was to design an innovative low-noise rail for high-speed trains. A slotted stand-off damping treatment was adopted for its effectiveness at reducing noise emission over a wide frequency range of 0–4000 Hz and for the possibility of applying it to existing rails. The choice of materials and treatment was the topic of the study. Several limitations had to be considered: ease of construction, shaping and assembly, safety, commercial availability, weight, and costs. To better understand the influence of the design parameters and to aid in the selection of those parameters, a numerical finite element calculation method for the loss factor was developed, and extensive studies on an actual rail model were conducted. The results of these analyses led to the construction of a prototype with a 2 mm thick steel constraining layer, 6 mm thick Dyad606 stand-off layer with twelve slots that were 3 mm deep and 20 mm wide and were equally spaced at intervals of 30 mm, and a viscoelastic layer of varying thickness according to the thickness of the rail waist and rail foot, arranged on both sides of the rail in the zone under the rail head.

Laboratory measurements of the accelerance of the samples confirmed that the slotted stand-off damping treatment shows significant effects on decreasing the amplitude of the accelerance of the rail in the frequency range of 0–4000 Hz, especially in the middle frequency range of 200–1000 Hz and the high frequency range of 1000–4000 Hz. Noise emission measurements in the frequency range of major acoustic interest (0–4000 Hz) were also carried out in the laboratory. Compared to the standard rail emission, good noise reductions of approximately 8.2 dB under vertical excitation and 9.4 dB under lateral excitation were achieved in this test. In particular, the noise level reduction in the frequency range of 200–4000 Hz was greater than that within the frequency range of 0–200 Hz under both vertical and lateral falling weight excitation. Following these encouraging results, low-noise wheels proceeded to the production stage.

ACKNOWLEDGMENT

This research was sponsored by the National Natural Science Foundation of China, no. U1234201. This support is gratefully acknowledged. The results and opinions presented are those of the authors and do not necessarily reflect those of the sponsoring agencies.

REFERENCES

1. C. Talotte, P.-E. Gautier, D. J. Thompson, and C. Hanson, "Identification, modelling and reduction potential of railway noise sources: a critical

survey," Journal of Sound and Vibration, vol. 267, no. 3, pp. 447–468, 2003.

2. M. Szwarc, B. Kostek, J. Kotus, M. Szczodrak, and A. Czyzewski, "Problems of railway noise—a case study," The International Journal of Occupational Safety and Ergonomics, vol. 17, no. 3, pp. 309–325, 2011.

3. J. J. Kalker and F. Périard, "Wheel-rail noise: impact, random, corrugation and tonal noise," Wear, vol. 191, no. 1-2, pp. 184–187, 1996.

4. C. Collette, "Importance of the wheel vertical dynamics in the squeal noise mechanism on a scaled test bench," Shock and Vibration, vol. 19, no. 2, pp. 141–149, 2012.

5. D. Thompson and C. Jones, "Noise and vibration from railway vehicles," in Handbook of Railway Vehicle Dynamics, pp. 211–345, CRC Press, London, UK, 2006.

6. D. T. Eadie, M. Santoro, and J. Kalousek, "Railway noise and the effect of top of rail liquid friction modifiers: changes in sound and vibration spectral distributions in curves," Wear, vol. 258, no. 7-8, pp. 1148–1155, 2005.

7. D. Thompson, "Wheel-rail noise generation. Parts I to V," Journal of Sound and Vibration, vol. 161, no. 3, pp. 387–482, 1997.

8. D. J. Thompson, "Experimental analysis of wave propagation in railway tracks," Journal of Sound and Vibration, vol. 203, no. 5, pp. 867–888, 1997.

9. J. Maes and H. Sol, "A double tuned rail damper—increased damping at the two first pinned-pinned frequencies," Journal of Sound and Vibration, vol. 267, no. 3, pp. 721–737, 2003.

10. W. Ho, B. Wong, and D. England, "Tuned mass damper for rail noise control," in Proceedings of the 10th International Workshop on Railway Noise, Noise and Vibration Mitigation for Rail Transportation Systems, Nagahama, Japan, 2010.

11. F. Ltourneaux, F. Margiocchi, and F. Poisson, "Complete assessment of rail absorber performances on an operated track in France," in Proceedings of the World Congress on Railway Research, CDROM, Montreal, Canada, 2006.

12. B. Asmussen, D. Stiebel, P. Kitson, D. Farrington, and D. Benton, "Reducing the noise emission by increasing the damping of the rail: results of a field test," Journal of Sound and Vibration, vol. 227, no. 1, pp. 711–721, 2005.

13. P.-B. Wei, H. Xia, Y.-M. Cao, and J.-W. Zhan, "Experimental study on vibration reduction of rail with compound damping board," Journal of Beijing Jiaotong University, vol. 31, no. 4, pp. 35–39, 2007 (Chinese).

14. Z. W. Li, X. Y. Lei, and P. F. Zhang, "FEM analysis of damped rails for vibration reduction," Noise and Vibration Control, vol. 74, pp. 64–66, 2009 (Chinese).

15. Y. I. N. Xue Jun, "Theoretical and experimental study on the labyrinth constrained damping layer,"Noise and Vibration Control, vol. 58, pp. 148–152, 2007 (Chinese).

16. P. Grootenhuis, "The control of vibrations with viscoelastic materials," Journal of Sound and Vibration, vol. 11, no. 4, pp. 421–433, 1970.

17. J. M. Yellin, I. Y. Shen, P. G. Reinhall, and P. Y. H. Huang, "An analytical and experimental analysis for a one-dimensional passive stand-off layer damping treatment," Journal of Vibration and Acoustics, vol. 122, no. 4, pp. 440–447, 2000.

18. J. M. H. Yellin, I. Y. Shen, and P. G. Reinhall, "Experimental and finite element analysis of stand-off layer damping treatments for beams," in Proceedings of the ASME International Mechanical Engineering Congress and Exposition (IMECE ‹07›), pp. 155–164, Seattle, Wash, USA, November 2007.

19. W. Wei, "A model of rail track reacceptance analysis," Journal of Dalian Railway Institute, vol. 19, no. 4, pp. 33–38, 1998 (Chinese).

20. International Organization for Standardization, "Vibration and shock-experimental determination of mechanical mobility—part 5: measurements using impact excitation with an exciter which is not attached to the structure," Tech. Rep. ISO 7626-5, ISO, Geneva, Switzerland, 1994.

21. International Organization for Standardization, "Acoustics-determination of sound power levels and sound energy levels of noise sources using sound pressure-engineering methods for an essentially free field over a reflecting plane," ISO 3744-2010, 2010.

Chapter 4

STUDY ON NOISE PREDICTION AND REDUCTION IN COUPLED WORKSHOPS USING SEA METHOD

Ye Lei[1,2], Jie Pan,[1] and Meiping Sheng[2]

[1]School of Mechanical and Chemical Engineering, The University of Western Australia, Nedlands, Perth, WA 6009, Australia

[2]School of Marine Engineering, Northwestern Polytechnical University, Xi'an 710072, China

ABSTRACT

A theoretical model for predicting noise reduction in coupled workshops is presented by using statistical energy analysis (SEA) method. An opening between the coupled workshops is considered into the theoretical model properly. The leakage issue is dealt with in the process of SEA modeling. An experiment is also carried out. A reasonable agreement between the prediction of noise reduction and the experimental data is observed. Moreover, it is concluded from the simulations that the sound energy transmit through the opening was the most important way to affect the noise reduction and the leakage is a significant element to influence the effect of noise treatment.

STATEMENT OF THE PROBLEM

The noise exposure standard of workplace in Australia is that the average exposure is 85 dB(A) in the form of overall sound pressure level over an eight-hour working day [1]. This regulation does not mean that below 85 dB(A) is safe, but that an eight-hour exposure of 85 dB(A) is considered to represent an acceptable level of risk to human hearing health. It is stated that if the noise level exceeds the regulation, action must be taken immediately. The investigated object in this paper is large coupled workshops, having a pump in one of them used for providing motility for water circulation, a double configured plasterboard placed in the middle of the workshops, and an opening beyond the plasterboard. When the pump runs, people in the receiving room feel so noisy that they cannot work efficiently and comfortably. Therefore, acoustical responses in both workshops and noise reduction between them should be assessed. If the overall sound pressure level exceeds 85 dB(A), noise treatment either on the noisy source, or on the propagation route should be carried out. In order to guide the noise treatment work, the effect of noise reduction after control also needs to be estimated. The configuration of the coupled workshops is shown in Figure 1. The left subfigure is the coupled workshops. The length of both room volumes is 15 m. The right subfigure describes the detailed configuration in the middle of coupled workshops, including an opening and double plasterboards. The plasterboards are fixed by screw and metal stud frame. Some rayon material fills the air gap between the plasterboards.

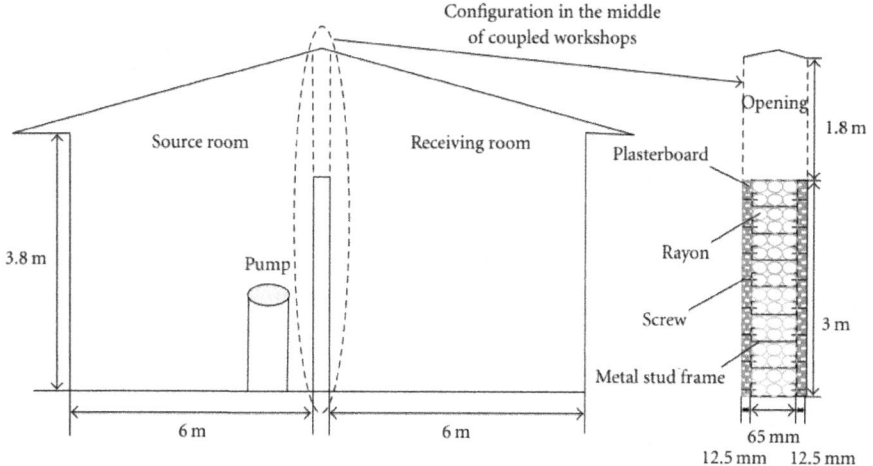

Figure 1: The configuration of the coupled workshops.

LITERATURE REVIEW

In practical project, various requirements need to be considered in the process of constructing buildings, thereby, no architecture can be built perfectly. Under the desire of better habitation and working environment, buildings are required to be constructed as ideally as possible. Therefore, investigation on acoustical properties of buildings is a meaningful topic, and the interrelated research has been developed for many years.

Furukawa et al. [2] proposed a two-dimensional SEA model for analyzing the characteristics of structure-born sound in a slender building and obtained the sound attenuation per unity distance. Wentang and Attenborough [3] also predicted sound pressure levels in coupled rooms using SEA method and the results are closer to the measured results than those obtained by the statistical acoustical method. In their study, the coupling between the window, doors, and walls is ignored, because the walls are so thick that most of the energy is dissipated within them and little is reradiated to other subsystems. In reality, leakage often exists in walls and windows, and many scholars found that it is crucial to transfer sound energy. Early studies by Sauter and Soroka [4] developed a theoretical model for two reverberant rooms with a slot between them. The slot was treated as a rigid rectangular piston while using the impedance function to express the relationship of sound pressures between the room volumes and slot, and only plane wave propagation was assumed through the slot. More recently, Billon et al. [5] analyzed coupled rooms with an aperture in them and proposed a diffusion model used for predicting acoustical responses taking the location into account, and the calculated results agreed well with the experimental data and those gained by the ray method. Franck et al. [6] gave a detailed review of methods for modeling apertures and presented a rigorous model for predicting the transmission loss through the aperture. In the model of Franck et al., propagating and evanescent acoustic modes were considered inside the aperture, and modal radiation impedance matrix was coupled with these modes in terms of modal contribution factors.

The aims of this paper are to take an opening into SEA model properly, to analyze the effect of leakage based on the model suggested by Gomperts and Kihlman [7], and to give constructive guidance for noise control. Section 3 presents the theoretical SEA models before and after noise control; Section 4 discusses the experimental studies, including measuring the sound pressure levels and reverberation times; Section 5addresses the comparison of the predicted and measured results and analyzes the effect of opening and leakage on energy transfer paths; Section 6 summarizes this research.

THEORETICAL ANALYSIS

SEA Model

In the process of building SEA model, it is complicated and of no need to express the whole vibroacoustic system exactly because of the basic statistical principle. For simplicity, it is assumed that there is no coupling between walls, floors, doors, windows, and ceilings. There are some equipments and furniture in the workshops. They provide necessary scattering conditions to satisfy statistical principle and are out of consideration in the SEA model.

Based on the above assumptions, the power flow relationships of subsystems are shown in Figure 2, where $\Pi_{i,\text{inis}}$ the input power to subsystem i, $\Pi_{i,\text{diss}}$ is the dissipated power in subsystem i, and $\Pi_{i,j}$ is the power exchanged between subsystem i and subsystem j.

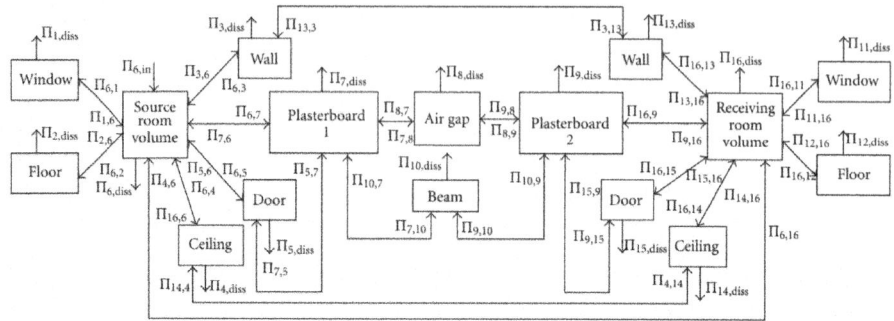

Figure 2: The energy flow relationships between subsystems before noise control.

A room system is divided into six kinds of sub-systems, including the room volume, the walls, the doors, the floor, the ceiling, and the windows. The power from the pump is inputted into the source room volume. The walls between the source room volume and the receiving room volume have direct coupling. The same coupling between ceilings exists. The door subsystem has coupling with the plasterboards. The path of transferring sound energy through the opening is represented as the direct coupling between the source room volume and the receiving room volume. Different from in the conventional SEA method, in the improved SEA model the indirect coupling between the room volumes and air gap inside the double plasterboards is ignored, but the nonresonant vibration and transmission of plasterboards [8] are included. The metal stud frame connecting two plasterboards is treated as beam class subsystems.

The basic equation for expressing the energy flow relationships between the ith subsystem and other subsystems in SEA model is

$$\omega \eta_i E_i - \omega \sum_{\substack{j \neq i}}^{N} \eta_{j,i} E_j = \Pi_i, \tag{1}$$

where the subscripts i and j, represent the subsystem i and j respectively, ω is the angle frequency, η_i is the DLF, $\eta_{i,i}$ is the coupling loss factor (CLF), E_i is the average modal energy, Π_i is the input power to subsystem i from outside environment, and N is the total number of subsystems. If there is no coupling between two subsystems, the CLF in (1) will be set to zero. The input power equals zero when there is no direct power from outside excitation.

The structural–structural CLF is calculated by [9]

$$\eta_{si,sj} = \frac{L_{si,sj} C_{g,si}}{\pi \omega S_{si}} \tau_{si,sj}, \tag{2}$$

where $L_{si,sj}$ is the junction length, $\tau_{si,sj}$ is the power transmission coefficient through the junction connecting structural subsystem i and j, and $C_{g,si}$ and S_{si} are the group velocity and area of structural element i.

The structural–acoustical and acoustical–structural CLFs are computed by [10]

$$\eta_{si,aj} = \frac{\rho_{aj} c_{aj}}{\omega \rho_{si}} \sigma_{si},$$

$$\eta_{aj,si} = \frac{\rho_{aj} c_{aj}^2 S_{si} f_{c,si}}{8\pi V_{aj} \rho_{si} f^3} \sigma_{si}, \tag{3}$$

where ρ_{aj}, c_{aj}, and V_{aj} are the medium density, sound velocity, and volume of room space j, respectively, and ρ_{si}, $f_{c,si}$, and σ_{si} are the surface density, critical frequency, and radiation efficiency [8, 11] of substructure i.

The acoustical–acoustical CLF is simulated as follows [9]

$$\eta_{ai,aj} = \frac{c_{ai} S_p}{4\omega V_{ai}} \tau_{ai,aj}, \tag{4}$$

where c_{ai} and V_{ai} are the sound velocity and volume, respectively, S_p is the area of structure used for separating room volumes, and $\tau_{ai,aj}$ is the transmission coefficient between the directly coupled room volumes. The transmission coefficient is supposed to be equal unity [6] for the case of opening and calculated by the model of Gomperts [7] for the case of leakage.

Finally, the average model energy of each subsystem can be obtained by solving (1). The squared sound pressure of acoustical subsystem is expressed by the average modal energy,

$$p_{ai}^2 = \frac{\rho_{ai} c_{ai}^2}{V_{ai}} E_{ai}. \tag{5}$$

The details of subsystems and energy flow relationships after treatment are shown in Figure 3. Different from in Figure 2, in Figure 3 the direct acoustical–acoustical coupling between the two room volumes is ignored, in the assumption that the craftwork is so consummate that no leakage existed. The opening is filled with double plasterboards, whose material and configuration are the same as the existing one. The nominations are accordant with those in Figure 2.

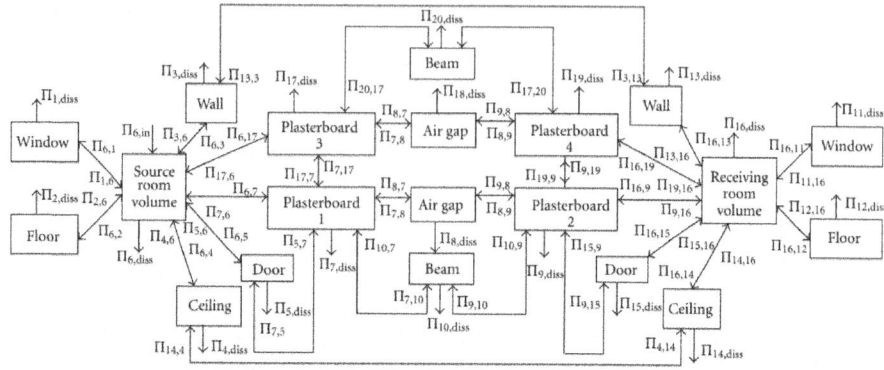

Figure 3: The energy flow relationships between subsystems after noise control.

Contribution Analysis

In order to guide the control work, investigation on energy transfer path is necessary. The contribution factor of the energy transfer path is defined as the ability of transferring energy and related with DLF and CLF of subsystems taking part in corresponding path. For an interesting subsystem, the total contribution from all the energy transfer paths is defined as a coefficient D, which is the average model energy ratio of the interesting subsystem and the source subsystem that is excited by external power directly

$$D = \frac{E_r}{E_s} = \frac{E_r^1 + E_r^2 + \cdots + E_r^i}{E_s}, \tag{6}$$

where E_r and E_s are the average model energy of the interesting subsystem and the source subsystem, respectively, and E_r^i represents the contributed energy of the ith energy transfer path to the overall energy of the interesting subsystem.

Separating (6) out, the contribution of each energy transfer path, also named the path loss factor [12], is described as the following equation:

$$D^i = \frac{E_r^i}{E_s}. \tag{7}$$

For example, there are n subsystems taking part in the route of the ith energy transfer path. From the source subsystem to the interesting subsystem, these acting subsystems are connected in the form of exchanging energy and supposed to be coupled one by one without the situation of one subsystem couples with two subsystems. Actually, one subsystem often couples with several other subsystems. If a subsystem couples with other two subsystems, the situation will be treated as two different paths respectively. Taking the first energy transfer path as a example, assuming there are three subsystems in the middle of this transmission path, denoting these three subsystems as the dth, eth, and fth subsystems, then the path loss factor is computed as

$$D^1 = \frac{E_r^1}{E_s} = \frac{\eta_{s\_d} \cdot \eta_{d\_e} \cdot \eta_{e\_f} \cdot \eta_{f\_r}}{\eta_d \cdot \eta_e \cdot \eta_f \cdot \eta_r}, \tag{8}$$

where η_{s_d} is the CLF between the source subsystem and the dth subsystem, and the physical meanings are analogically for η_{d_e}, η_{e_f}, and η_{f_r}. Symbols η_d, η_e, η_f, and η_r are the DLFs of subsystem d, e, f and the interesting subsystem, respectively.

Equation (7) is divided by (6), and the percentile of each contributor can be obtained

$$P^i = \frac{D^i}{D} = \frac{D^i}{\sum_i D^i}. \tag{9}$$

EXPERIMENTAL STUDIES

In order to understand the acoustical properties of the pump and obtain data for calculation and validation, some experimental work was finished firstly before noise control. The motor of the pump runs in a rotor speed around 2900 r/min, which is the normal used speed but not the maximal one. The measurement locations are distributed symmetrically in the center line of the source room and receiving room respectively. The measured average sound pressure levels are shown in Figure 4.

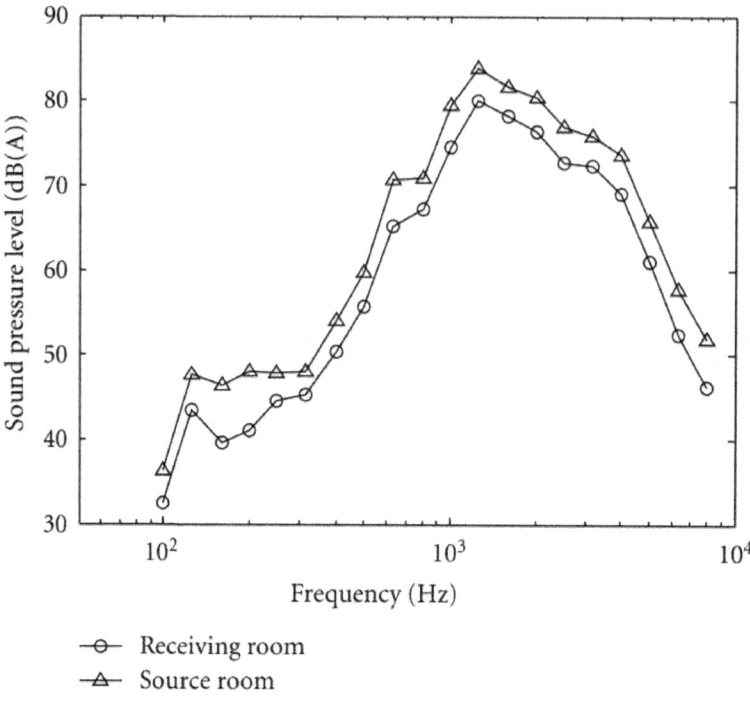

Figure 4: Measured sound pressure levels before noise control.

From the curves in Figure 4, it can be found that the maximum response in both rooms is around 1.25 kHz, which is the harmonic frequency of the rotor. This frequency is in the sensitive frequency range of human hearing, around 1 kHz. Consequently, a person in the receiving room will feel very uncomfortable if the pump works.

Furthermore, attention should be paid to the overall sound pressure level, which is 88.7 dB(A) in the source room and 84.7 dB(A) in the receiving room, respectively. Unfortunately, the sound pressure level in the source room exceeds the regulation of 85 dB(A). However, the value in the receiving room just arrives the limitation, but it will also exceed 85 dB(A) if the pump runs in a speed which is a little larger than 2900 r/m.

From the above analysis of measured results in coupled workshops, it is required that noise treatment should take place. The ascendant decision is to fill the opening. A single side of the double plasterboards used for filling the opening is shown in Figure 5. The thickness is 12.5 mm.

Figure 5: The plasterboard filling the opening used for noise control.

Similarly, the measurement of acoustical responses was also carried out after noise treatment, and the sound pressure levels are plotted in Figure 6.

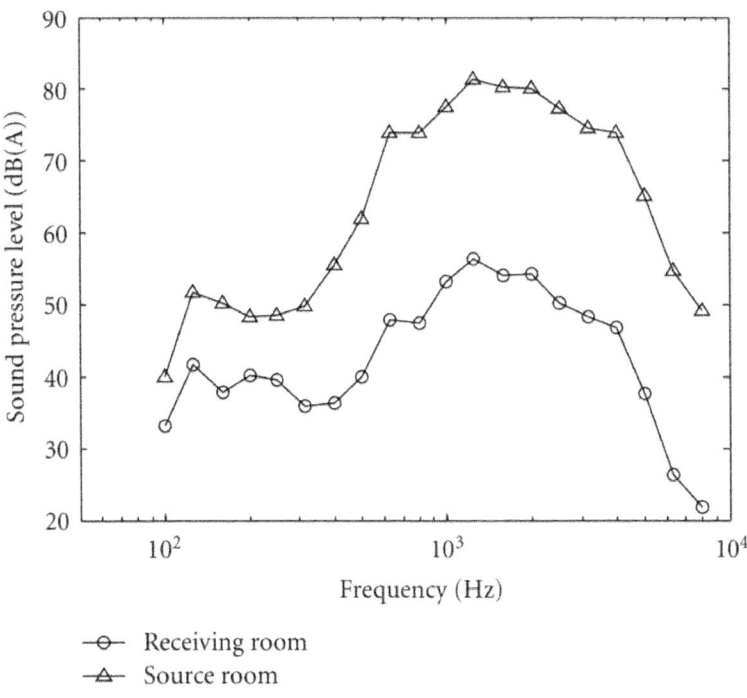

Figure 6: Measured sound pressure levels after noise control.

Damping loss factor (DLF) is one of the basic parameters in SEA modeling and used for predicting the power dissipated inside the subsystem. Because it is unrealistic to measure the DLF of structural subsystem separately, approximate values of concrete, brickwork, plywood, and glass are cited from [13]. Jaime and Manuel [14] proposed a fitting formula for computing DLF of plasterboard, as follows:

$$\eta_{d\_plasterboard} = 0.4171 f^{-0.4467} \tag{10}$$

The DLF of room volumes can be calculated by measuring their reverberation time $T60$. The relationship between the DLF and the reverberation time $T60$ can be obtained from the half-power bandwidth method and is expressed as follows:

$$\eta_d = \frac{2.2}{f T_{60}} \tag{11}$$

The measured DLFs of the room volumes before and after noise control are shown in Figure 7. Because the source power is sufficient above 100 Hz, the measured results are reliable. It can be found that the DLFs of these two workshops have a great agreement above 400 Hz before control and have little difference at almost all frequencies after control.

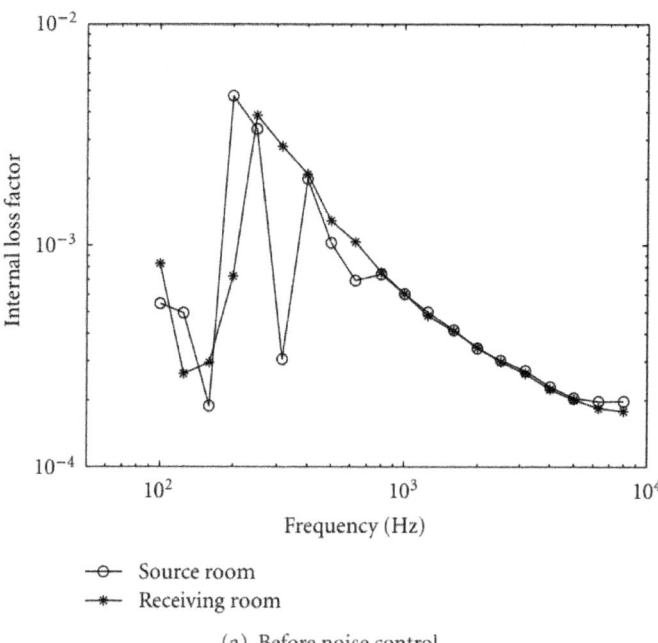

(a) Before noise control

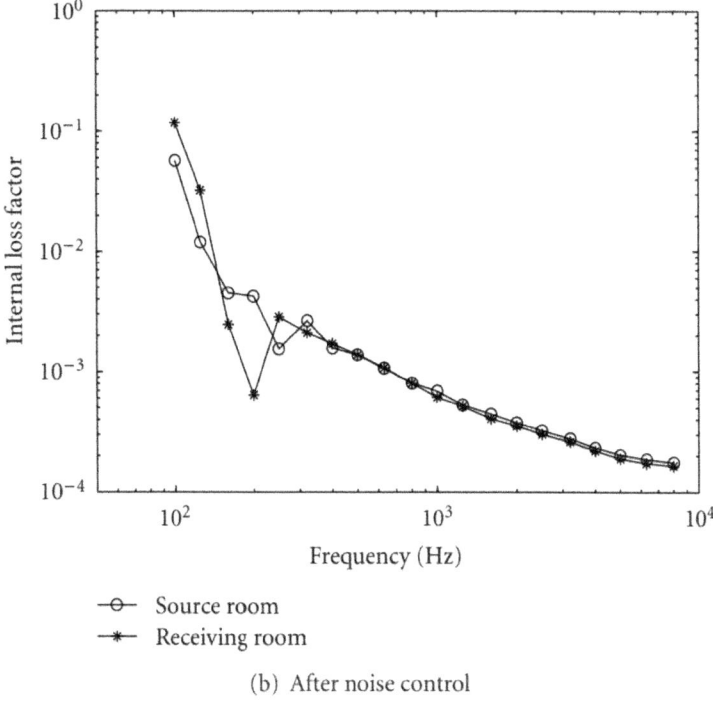

(b) After noise control

Figure 7: The DLFs of the coupled room volumes.

RESULTS AND DISCUSSION

Before Noise Control

The detailed values of material property and DLF are listed in Table 1.

Table 1: The material properties and DLFs of structures

Material	Density kg/m^3	Young's modulus Pa	Poisson's ratio	DLF
Concrete (floors)	1900	1.6×10^{10}	0.25	0.02
Brickwork (walls)	1900	1.6×10^{10}	0.2	0.02
Plasterboard (clapboard and ceiling)	800	2×10^9	0.26	Calculated by Jaime et al. [14]'s formula
Plywood (doors)	700	6×10^9	0.25	0.01
Glass (windows)	2300	6.2×10^{10}	0.24	0.01

Based on above analysis, using the acoustical response in the source room as the input condition for (1), the predicted sound pressure level difference between the coupled workshops is plotted in Figure 8. Error range of 3 dB is given in the figure for analyzing the deviation between the predicted and measured results. There are many causes for the prediction error. From [6], if the sound transmission coefficient through the opening equals unity, the maximum error is 2 dB. The measurement error is setting to 1 dB in sound pressure level meter (B&K2238) during experiment. Hence, the total prediction error is larger than 3 dB because of another undetermined error coming from hypotheses proposed in the foregoing section.

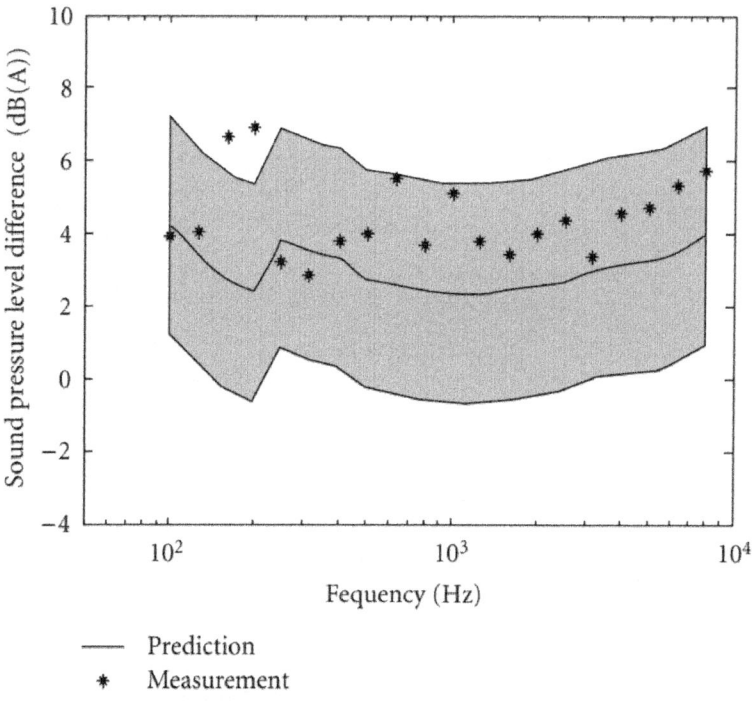

Figure 8: The sound pressure level difference before noise control.

From the results in Figure 8, it is found that the maximum noise reduction is nearly 7 dB(A) around 200 Hz and the average noise reduction is 4 dB(A) from 100 Hz to 8 kHz. It is noticeable that most of the measured results are in the error bar and the maximum deviation is approximately 5 dB(A) at 160 Hz and 200 Hz, so the theoretical model is reliable.

In addition, the contribution of the most important energy transfer path and other ones is plotted in Figure 9, where the quantity of y-axis is the contribution in the form of percentage. Figure 9(a) represents that the sound energy in the source room transmits through the opening to the receiving room, and Figure 9(b) is the contribution of other paths except that in Figure 9(a). It is shown that the most important energy transfer way is that the sound energy transmits through the opening directly, and the percentage is 15% at a low frequency of around 100 Hz and increases smoothly until 400 Hz to more than 95%, and the value has no evident changes except 3.15 kHz with the percent of 83%. The comparison between the most significant path and other ones indicated that the treatment should be focused on the opening and other contributors can be ignored, especially at the intermediate and high frequencies.

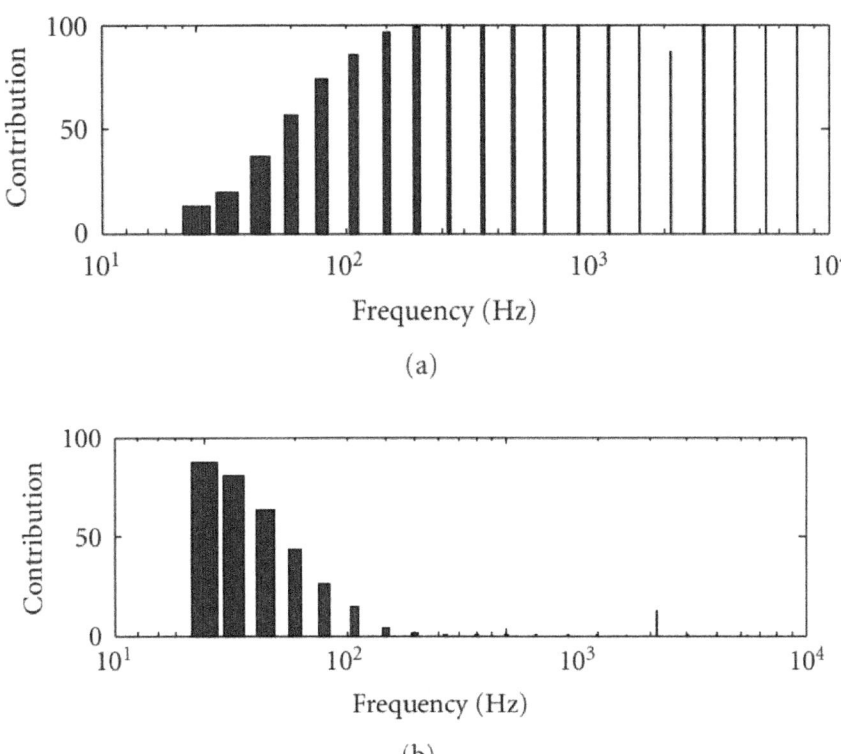

Figure 9: The contribution of energy transfer paths before noise control ((a): the energy transfer path from the source room through opening to the receiving room; (b): other energy transfer paths except subfigure (a)).

After Noise Control

From the prediction and analysis before noise control, it is affirmed that most of the sound energy in the noisy source room is transmitted to the receiving room through the opening and filling it is a straightforward way to control the noise level in the receiving room. But if leakage appears during the treatment work, the noise reduction will be affected remarkably [7]. Consequently, analyzing the noise reduction with a different leakage area is meaningful to practice. Figure 10 gives three cases of configuration of a different leakage area that will be discussed in the following content. The different area is described by different width and consistent length and depth. The length is 15 m and the depth is 0.09 m.

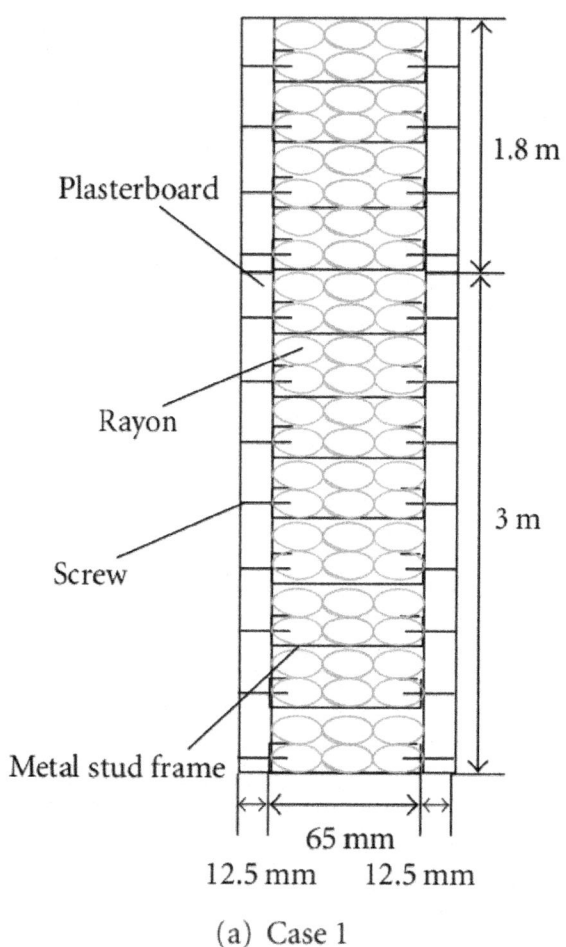

(a) Case 1

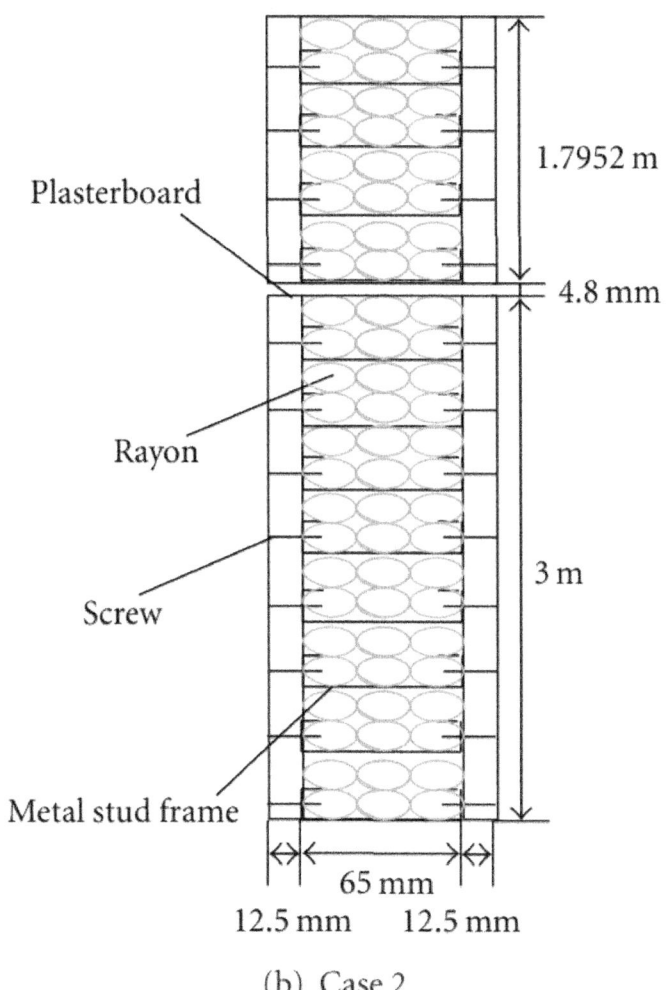

(b) Case 2

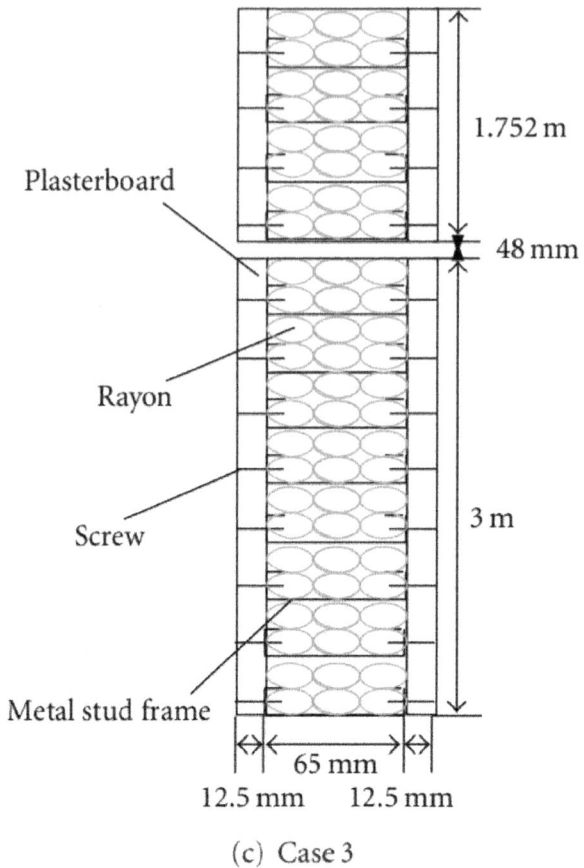

(c) Case 3

Figure 10: The plasterboard after noise control with a different area of leakage (Case 1: no leakage and the padding plasterboard connects with the existing one very well; Case 2: the area of the leakage occupies 0.1% of the total area of the plasterboards; Case 3: the area of the leakage occupies 1% of the total area of the plasterboards).

The predicted sound pressure level differences, corresponding to the cases in Figure 10, are shown in Figure 11. It is shown apparently that the leakage is of great importance to the noise reduction. When there is no leakage existing between the designed and existing plasterboards, the effect of noise treatment is wonderful above 630 Hz, and the average sound pressure level difference reaches 49 dB(A) in the whole interesting frequency domain. If there is 0.1% leakage, the noise reduction will decline from 400 Hz of approximately 23 dB(A) and reduce to approximately 20 dB(A) at 8 kHz, compared with the values of no leakage, and the average sound pressure level difference is 20.8 dB(A). When the area of leakage increases to 1% of the total area of the plasterboards, the noise reduction is about 5 dB(A) higher than the quantity of no treatment at low frequencies and is almost the same with the value of 0.1% leakage at high frequencies; and the average sound pressure level difference drops to 13.2 dB(A). It can be concluded that the area of leakage changing from zero to nonzero will affect the noise reduction at high frequencies more significantly and with the increase of the area of leakage, the influence of leak becomes more remarkable at low and intermediate frequencies.

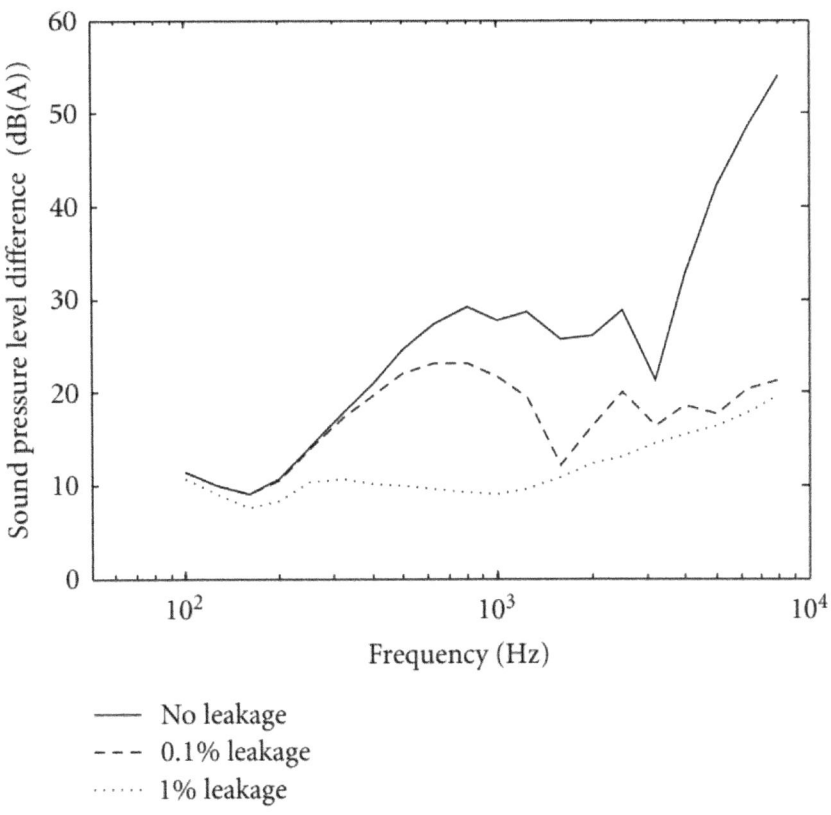

—— No leakage
--- 0.1% leakage
······ 1% leakage

Figure 11: The sound pressure level difference after noise control.

According to the conclusion from Figure 11, it is suggested that the padding plasterboard should connect with the existing one as the more sealed, the better. Through observing the accomplished controlling work, there is no obvious leakage existing. Therefore, the SEA model with no leakage was used to predict the noise reduction after control. The measured and predicted sound pressure level difference is drawn in Figure 12. The highest analyzing frequency is 4 kHz, because of the insufficient signal noise ratio, which is different from the results before control.

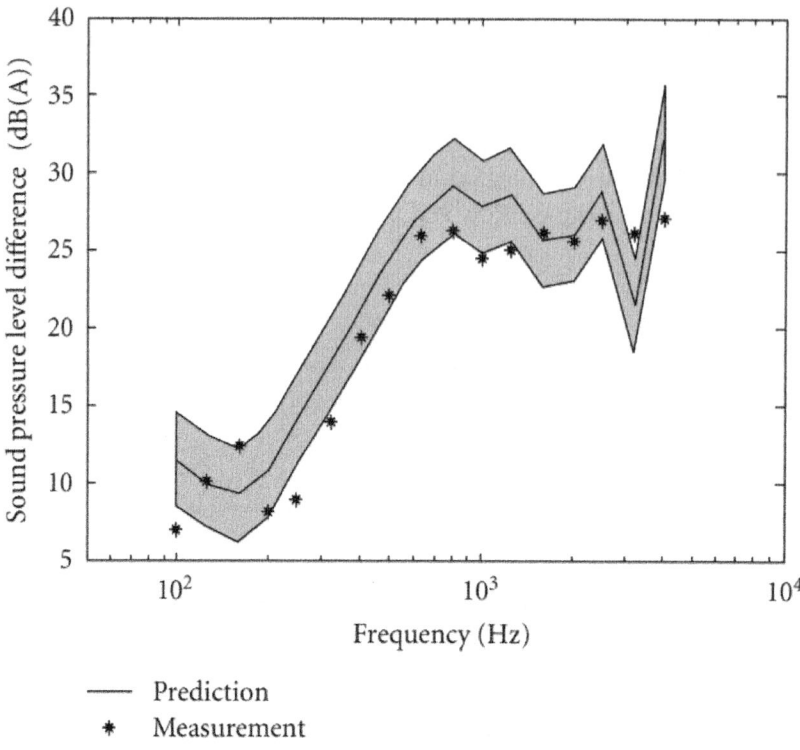

Figure 12: The sound pressure level difference after noise control.

The average noise reduction is 26.7 dB(A) and the maximum reduction of noise is about 30 dB(A). Almost all measured data drops in the error range of 3 dB, so the reliability of theoretical model is validated again.

CONCLUSIONS

Some research on predicting noise reduction and controlling noise in coupled workshops was finished by using SEA method, including the cases of considering opening, leakage, and perfect situation without sound leak. The opening was taken as a filter in which the transmission coefficient between the coupled workshops through it equals unity. The leakage was taken into consideration by adopting the model of Gomperts et al..

The measured sound pressure level before noise control affirmed that the acoustical responses in both workshops exceed the noise exposure regulation of workplace and need to be controlled. Before treatment, most of the measured sound pressure level difference is in the error bar of prediction of 3 dB. The average noise reduction is 4 dB(A) and the most important way to transfer sound energy is through the opening directly and occupies more than 95% of the total contributors above 400 Hz.

After noise control, the measured results also have good agreement with the prediction. By comparing the noise reduction before and after control, the sound pressure level difference increases by 22.7 dB(A) in the form of overall sound pressure level. It is also concluded that leakage is significant to noise reduction. When the leakage changes from none to existing, the effect of decreasing noise reduction is more observable at high frequencies, and with the increment of the leakage area, the noise reduction at low and intermediate frequencies is affected more remarkably.

ACKNOWLEDGMENTS

This research has been undertaken in the University of Western Australia (UWA), under the joined PhD program between the Northwestern Polytechnical Universtiy (NWPU). Meanwhile, the experiment work is finished in the School of Civil and Mechanical Engineering in UWA and has received support from Mr. Rob Greenhalgh for organizing and arrangement. The group of Professor Krish Thiagarajan, Dr. Hongmei Shun, Dr. Yanni Zhang, and Dr. Ye Zhihui also provided assistance for the measurement work. The financial supports from UWA and the joined program and all the above-mentioned people are gratefully acknowledged.

REFERENCES

1. "National Code of Practice for Occupational Noise NOHSC:2009," Published by the National Occupational Health and Safety Commission, 2004.
2. H. Furukawa, K. Fujiwara, Y. Ando, and Z. I. Maekawa, "Analysis of the structure-borne sound in an existing building by the SEA method," Applied Acoustics, vol. 29, no. 4, pp. 255–271, 1990. ·
3. R. Wentang and K. Attenborough, "Prediction of sound fields in rooms using statistical energy analysis," Applied Acoustics, vol. 34, no. 3, pp. 207–220, 1991. ·
4. A. Sauter Jr. and W. W. Soroka, "Sound transmission through rectangular slots of finite depth between reverberant rooms," Journal of the Acoustic Society of America, vol. 47, no. 1A, pp. 5–11, 1969.
5. A. Billon, V. Valeau, A. Sakout, and J. Picaut, "On the use of a diffusion model for acoustically coupled rooms," Journal of the Acoustical Society of America, vol. 120, no. 4, pp. 2043–2054, 2006. · ·

6. S. Franck, N. Hugues, and A. Nouredddine, "On the modeling of the diffuse field sound transmission loss of finite thickness apertures," Journal of the Acoustical Society of America, vol. 122, no. 1, pp. 302–313, 2007. · ·
7. M. C. Gomperts and T. Kihlman, "The sound transmission loss of circular and slit-shaped aperture in walls," Acustica, vol. 18, pp. 144–150, 1967.
8. Y. Lei, J. Pan, and M. P. Sheng, "Noise reduction of an acoustical enclosure—mechanisms and prediction accuracy," in Proceedings of 20th International Congress on Acoustics, Sydney, Australia, August 2010.
9. R. H. Lyon, Statistical Energy Analysis of Dynamical Systems: Theory and Applications, M.I.T. Press, Cambridge, Mass, USA, 1975.
10. T. R. T. Nightingale and I. Bosmans, "Expressions for first-order flanking paths in homogeneous isotropic and lightly damped buildings," Acta Acustica, vol. 89, no. 1, pp. 110–122, 2003. ·
11. I. L. Ver and C. I. Holmer, "Interaction of sound waves with solid structures," in Noise and Vibration Control, L. L. Beranek, Ed., McGraw-Hill, New York, NY, USA, 1971.
12. A. T. Chavan and D. N. Manik, "Design sensitivity analysis of statistical energy analysis models using transfer path approach," Electronic Journal Technical Acoustics, vol. 3, 2005.
13. M. Dayou, Handbook of Noise and Vibration Controlling Engineering, China Machine Press, Beijing, China, 2002.
14. D. Jaime and R. Manuel, "Loss factor measurements on plasterboard," in Proceedings of 19th International Congress on Acoustics, Madrid, Spain, September 2007.

Chapter 5

INFLUENCE OF SOUND VIBRATION ON DIAMOND-LIKE CARBON DEPOSITION RATE

Syed Md. Ihsanul Karim,[1] Mohammad Asaduzzaman Chowdhury,[2] and Md. Maksud Helali[3]

[1]Bangladesh Industrial Technical Assistance Centre (BITAC), Ministry of Industries, Dhaka 1208, Bangladesh

[2]Department of Mechanical Engineering, Dhaka University of Engineering and Technology, Gazipur 1700, Bangladesh

[3]Department of Mechanical Engineering, Bangladesh University of Engineering and Technology, Dhaka 1000, Bangladesh

ABSTRACT

This work examines how vapor-deposited coating of DLC (partially diamond) on stainless steel 304 substrate is affected by the sound vibration. For this, a specially designed chemical vapor deposition (thermal CVD and hot filament) apparatus having facility of generating sound vibration at different frequency is fabricated. A coating of DLC (partially diamond) has been deposited on the substrate, and the characterization of the coating has been done by SEM, EDX, and XRD. The coating of carbon is identified by EDX, and the allotropic forms of graphite and diamond peaks of carbon are found by XRD analysis. By SEM analysis, it is found that the microstructures of deposited coatings are more compact and smoother under vibration than those in absence of vibration. The experiments were conducted under different ranges of vibration including sonic and ultrasonic range. Studies have shown that the growth rate of deposited coating on a unit area is higher under vibration than that in absence of vibration. It is found that deposition rate varies with the distance between substrate and activation heater and frequency of vibration. The deposition rate does not vary significantly with the change of frequency in the sonic range. The amount of deposition under ultrasonic vibration increases significantly with the frequency of vibration upto 5-6 mm distance between substrate and activation heater. Within this distance, the difference of deposition rate under vibration and without vibration conditions increases almost linearly with the increase of frequency of vibration. Beyond this distance, the effect of frequency on deposition rate becomes almost constant. In addition, the higher the distance, the less is the effectiveness of frequency of vibration on the deposition rate in that range. The deposition rate increases due to the extra vibration of sound added to the system which may enhance the activation energy by increasing its kinetic energy. The experimental results are compared with those available in the literature, and physical explanations are provided.

INTRODUCTION

Chemical vapor deposition (CVD) is a process in which a solid material formed from a vapor phase by chemical reaction is deposited on a heated substrate. The deposited material is obtained as a coating of multicrystal layer. The controlling parameters in CVD process are surface kinetics, mass transport in the vapor, thermodynamics of the system, chemistry of the reaction and processing parameters like temperature and pressure. The deposition rate which is the prime limiting factor in a CVD process is mainly controlled by the formation of required species to be deposited and its transportation in the vapor and surface kinetics [1–3]. Several authors [4–7] observed that the quality and the rate of deposition depend on temperature of the substrate and filament, gas flow rate, gas composition (reactants), and chamber pressure. Different arrangements and techniques such as centrifugation, vertical vibration of the substrate, hydrogen and argon inclusion, change of gas injecting location, and formation of plasma have been studied by different authors [8–12]. The effects of operating parameters on the deposition rate were also investigated for these techniques. A study is carried out to observe the effect of ultrasonic vibration on electrochemical deposition [13]. However, the effect of sound vibration on CVD process is yet to be investigated. Kinetic energy increases by adding extra energy of sound which may increase the deposition rate. This extra kinetic energy may enhance the chemical activity by overcoming the potential barrier and increases the mass transport of the species.

The effect of sound vibration increases with the increase of density of media, through which it travels. Therefore, the CVD process has been selected, as CVD does not usually require very low pressure, which is necessary for PVD system. Consequently, the vacuum system in CVD is simpler and less costly.

Comparing with other CVD process, thermal CVD (hot filament) is relatively inexpensive, and experiments can be readily carried out. Therefore, in this study an attempt is made to investigate the effect of sound vibration, in particular, the frequency of vibration on the deposition rate. In addition to deposition rate, the quality of deposited coating is also investigated. Deposition in absence of vibration was investigated first, and then the results were compared with the results obtained under different frequency of vibrations. Some parameters that affect the deposition were also inspected.

EXPERIMENTAL

A thermal chemical vapor deposition (hot filament) setup (Figure 1) was designed and fabricated. The setup is a CVD system comprises of a reactor chamber supported by some subassemblies and sub systems. The subassemblies are

i. heater,
ii. sound generating system,
iii. connector and
iv. cooling line, and

the subsystems are

i. gas evacuation system,
ii. electric supply system,
iii. heating system,
iv. cooling system,
v. gas supplying system,
vi. substrate cleaning system
vii. measuring system, and
viii. structure and handling system.

These arrangements of the experimental setup (Figure 1) are similar to the conventional thermal chemical vapor deposition (hot filament) unit.

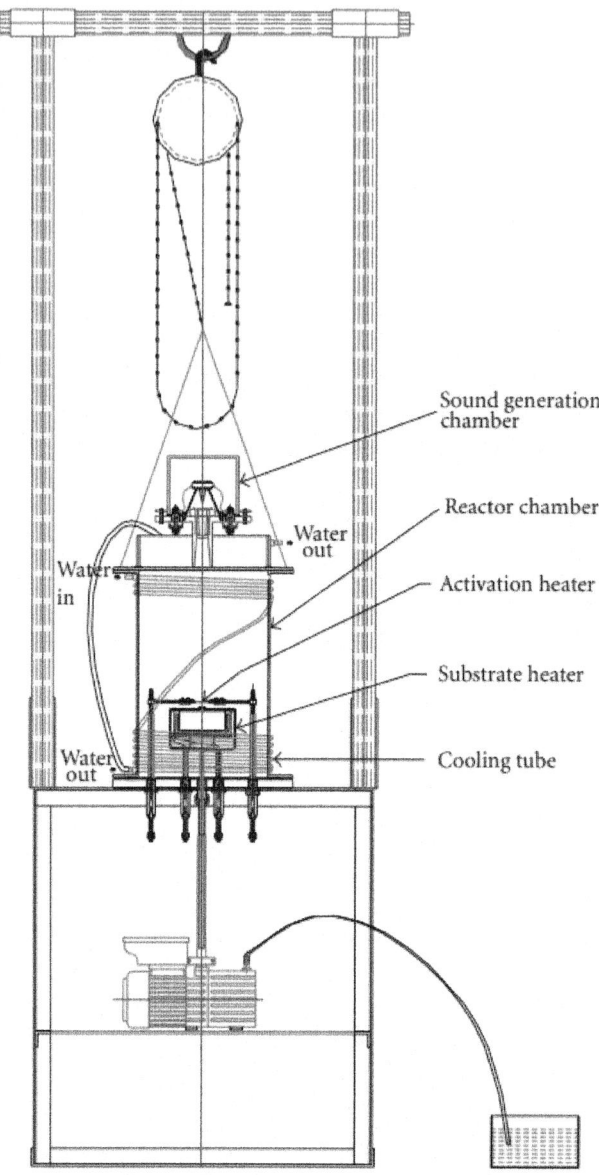

Figure 1: Schematic diagram of chemical vapor deposition (hot filament) setup.

A separate arrangement is designed and fabricated for generating sound shown, in Figure 2. The signal of sound vibration is generated by signal generator. After amplification by an amplifier, this signal of sound vibration passes through the wire to the piezoelectric horn, placed inside the vibration generating chamber. There are two insulated leak proof connectors in the vibration generating chamber, which facilitate to pass the sound signal from outside to inside. The sound is generated in the piezoelectric horn since it gets the sound generating signal from signal generator and passes through hollow pipe towards the substrate. There is a provision to monitor the frequency and amplitude of the generated sound vibration by an oscilloscope, connected parallel with the input wire.

(1) lower part
(2) rubber seal between upper part and lower part
(3) clamping bolt
(4) S. S. washer plate
(5) connector
(6) upper part
(7) horn plastic body
(8) connecting wire
(9) plastic dome-like body
(10) metallic connector
(11) horn head
(12) internally threaded hollow pipe
(13) rubber seal between bottom part and hollow pipe
(14) horn clamping bolt
(15) rubber seal between horn body and lower part
(16) rubber seal between bottom part and hollow bolt
(17) hollow bolt

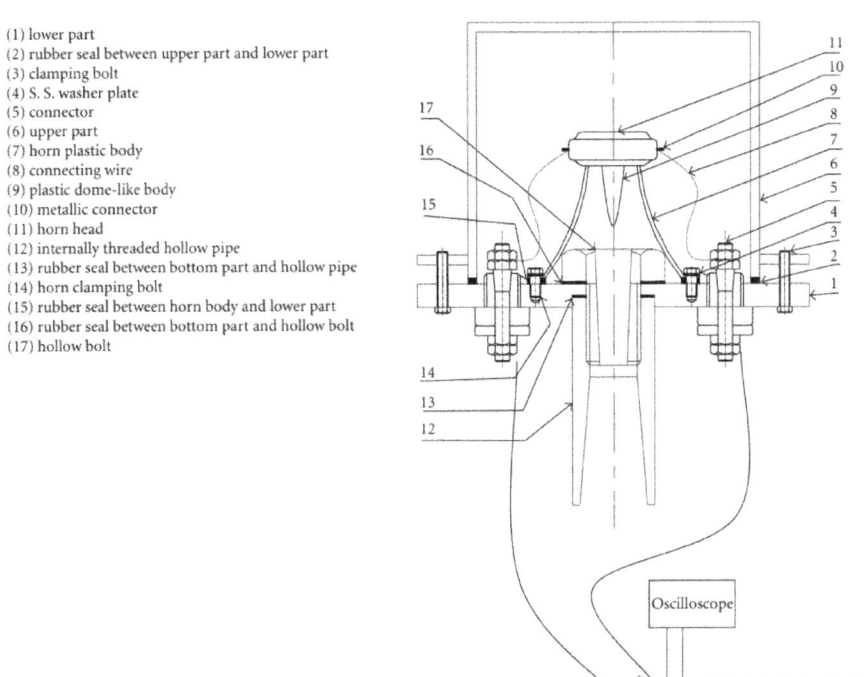

Figure 2: Schematic diagram of sound generation system.

The deposition rates of the coating per unit area per unit time were calculated from the weight difference of substrate before and after deposition. The surface morphologies of the deposited coatings were analyzed by SEM attached with energy dispersive X-ray spectrometry (EDX). X-ray diffraction (XRD), with target of Mo (Zr), 30 kV/20 mA, and an incident angle of 1°, is used to study the composition of coated material.

Experimental conditions are shown in Table 1. During tests, each experiment was repeated several times. In the figures, dispersion of test results at each point is also shown.

Table 1: Experimental variables

S. number	Parameters	Range
(1)	Pressure	20–30 Torr
(2)	Substrate (nicrom) heater temperature	800–1000°C
(3)	Activation (tungsten) heater temperature	1800–2000°C
(4)	Substrate (nicrom) heater power	1000 watt
(5)	Substrate (nicrom) heater voltage	80–100 V
(6)	Substrate (nicrom) heater current	7–10 amp
(7)	Activation (tungsten) heater power	200 watt
(8)	Activation (tungsten) heater voltage	5–7 V
(9)	Activation (tungsten) heater current	25–35 amp
(10)	Flow rate (CH_4 gas)	0.1–1.5 L/min
(11)	Gap between substrate and Tungsten heater	2.5–8.0 mm
(12)	Sound frequency	0–110 kHz
(13)	Deposition duration	3–10 minutes
(14)	Substrate size	22 mm × 14 mm × 1.15 mm

RESULTS AND DISCUSSIONS

Figure 3 shows the variation of deposition rate under vibration and without vibration in the range of frequency 0 to 110 KHz at a distance of 5 mm between substrate and activation heater. At each frequency, at least five experiments were conducted to get an average deposition rate. All these data are presented in the figure. From these data, a considerable increase in deposition rate on unit area is observed for deposition under vibration compared to that under no vibration.

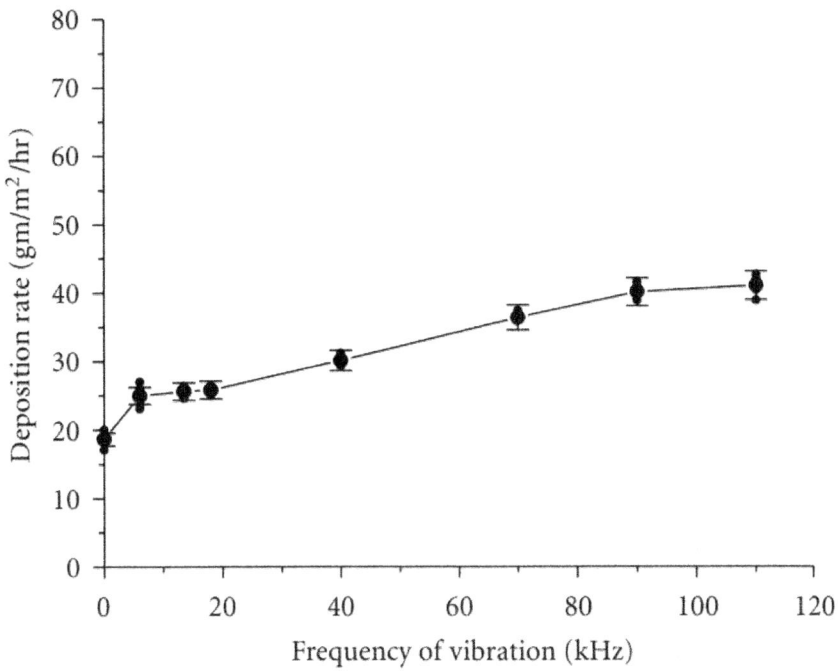

Figure 3: Deposition rate as a function of frequency of vibration (T_{sub} = 1000°C, T_{act} = 1900°C, P_{ch} = 2 5 Torr, and d = 5mm).

From this figure, it is observed that deposition rate increases from 0 at no to approximately 6 KHz. From 6 KHz, the deposition rate increases linearly up to approximately 20 KHz, and after that the steepness of the curve shows higher deposition rate, and finally increment rate reduces to almost negligible amount up to the observed range. The higher deposition rate under vibration might be due to the fact that mechanical and pressure wave of propagated sound towards the substrate enhances the mass transfer rate of depositing carbon species. The variation of the rate of increment of deposition at different frequency ranges might be due to the change of resultant vibration of carbon particles which depends on the wave length of the sound at different frequency and the particle size and the mass of the species. It is observed that deposition rate under sonic range of sound vibration does not vary significantly with frequency of vibration. But the rate of deposition under ultrasonic vibration increase significantly with frequency of vibration.

Figure 4 shows the variation of deposition rate with the distance between substrate and activation heater under vibration and no vibration. From this figure, it is observed that deposition rate under vibration is higher than that under no vibration at identical conditions within the observed range. Curves 1,

2, 3, and 4 of Figure 4 are drawn for 4 mm to 8 mm distance between substrate and activation heater and under the vibration of 0, 40, 70, 90, and 110 kHz, respectively. From this figure, it is shown that the deposition rate decreases linearly with the increase of distance between substrate and activation heater up to 6 mm, and after that it remains almost constant. This may be due to the numbers of activated carbon species and atomic hydrogen are more near the activation tungsten heater. It is observed that the decreasing deposition rate with distance (up to a certain distance) is higher for high frequency in comparison to lower frequency. This is because the higher the intensity, the higher is the variation of energy with respect to distance. For a particular distance, the deposition rate is high for higher frequency. This is because, at higher frequency, the intensity of sound is higher, and thus the energy transmitted from sound to the particle is higher. For all frequencies, when the distance is more than 5.5 mm the deposition rate is almost constant. That means, beyond this distance, the effect of frequency on deposition rate is negligible. This may be due to the dissipation of sound energy to the environment reaches to a mean value (for any intensity) at a distance more than 5.5 mm.

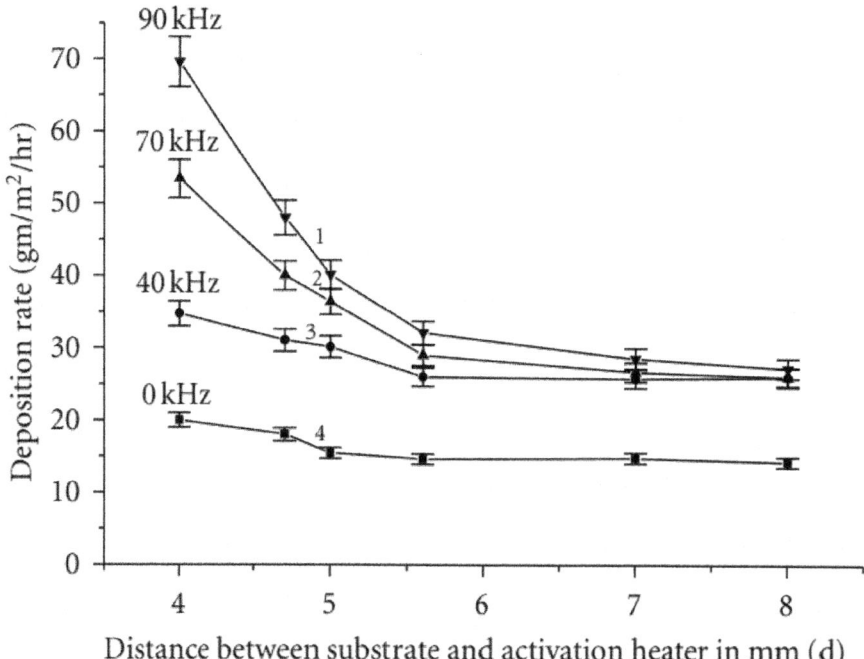

Figure 4: Effect of vibration on deposition rate with respect to distance between substrate and activation heater (T_{sub} = 1000°C, T_{act} = 1900°C, and P_{ch} = 25 Torr).

Figure 5 shows the comparison of micrographs of coated surface under vibration and without vibration at 200 magnifications. The left-side view of this figure is the coated surface deposited under vibration (40 kHz), and right-side view is the coated surface deposited without vibration. By observing the morphologies of the coated surfaces (Figure 5), it is shown that the deposited coating under vibration is more compact, smooth as compared to the coating deposited without vibration.

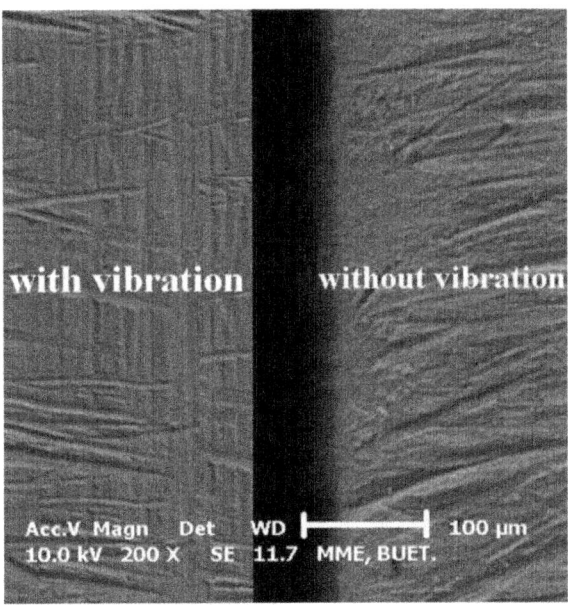

Figure 5: Microstructure (under SEM) of deposited coating under vibration (left-side view) and without vibration (right-side view) condition at different resolutions.

The EDX analysis shows that the coating on the substrate has considerable amount of carbon particles under vibration (13%) and without vibration (11%) condition as shown in Figure 6. This indicates that vibration of sound creates mechanical and pressure wave propagation, which influences the mass transport towards the substrate. As a result, better coating is obtained under sound vibration compared to that without vibration.

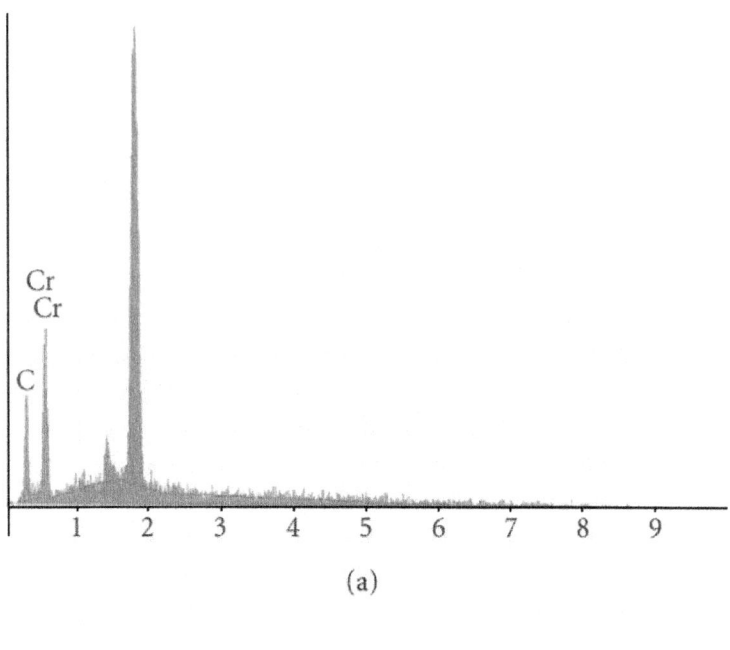

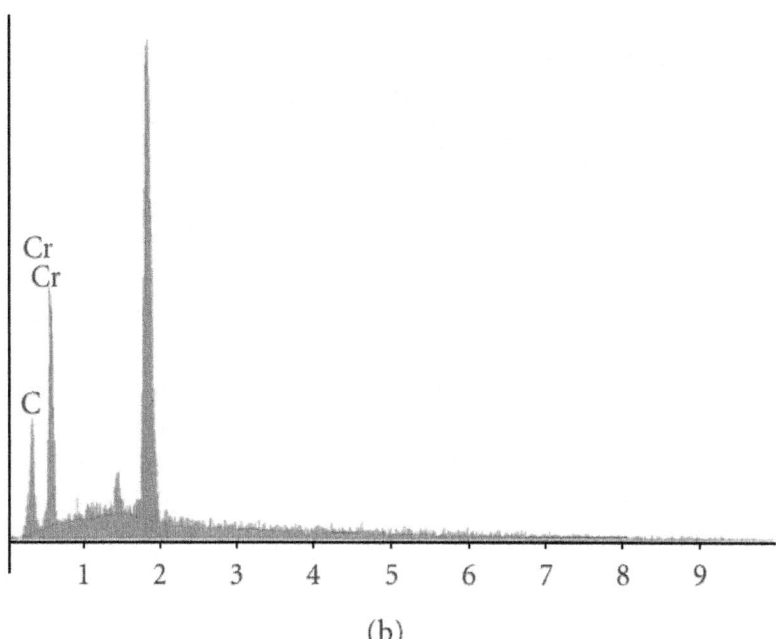

Figure 6: EDX analysis of coating on substrate with vibration at top and without vibration at bottom ($T\text{sub} = 1000°C$, $T_{act} = 1900°C$, $P_{ch} = 25$ torr, and $L_{gap} = 5$ mm).

Influence of Sound Vibration on Diamond-Like Carbon Deposition Rate

XRD analysis on specimens is shown in Figure 7 for base metal without coating, coating without sound vibration, and coating with sound vibration. The d-spacings determined from XRD patterns of the existing crystals within the coated layer without and with vibrations are shown in Tables 2 and 3 respectively. They are compared with the standard d-spacings for graphite, diamond, and austenite of austenitic stainless steel. It is found that the values of d-spacings calculated from the observed peaks and from XRD analysis are almost similar with the standard d-spacings values of graphite, diamond, and austenite of austenitic stainless steel.

Table 2: Comparison of the d-spacings of XRD spectrum of the deposited crystal with the actual d-spacings for graphite, diamond, and Fe (γ) without sound

Peak number	2θ	Measured d (A)	i/i_0	Diamond (h k l)	d (A)	Graphite (h k l)	d (A)	Fe (γ) (h k l)	d (A)
(1)	19.5	2.0982	89					111	2.095
(2)	21.2	1.9317	64	105	1.930	103	1.9200		
(3)	22.5	1.8214	86					200	1.8214
(4)	23.3	1.7597	53	108	1.6650	104	1.7950		
(5)	26.0	1.5796	14	109	1.5800	106	1.5400		
(6)	32.6	1.2660	100					220	1.283
(7)	33.4	1.2365	14	101 4	1.2200	110	1.2280		

Table 3: Comparison of the d-spacings of XRD spectrum of the deposited crystal with the actual d-spacings for graphite, diamond, and Fe (γ) with sound

Peak number	2θ	Measured d (A)	i/i_0	Diamond (h k l)	d (A)	Graphite (h k l)	d (A)	Fe (γ) (h k l)	d (A)
(1)	19.5	2.0982	92					111	2.095
(2)	21.2	1.9317	42	105	1.930	103	1.9200		
(3)	22.5	1.8214	100					200	1.8214
(4)	23.3	1.7597	83	108	1.6650	104	1.7950		
(5)	26.0	1.5796	17	109	1.5800	106	1.5400		
(6)	32.6	1.2660	75					220	1.283
(7)	33.4	1.2365	21	101 4	1.2200	110	1.2280		

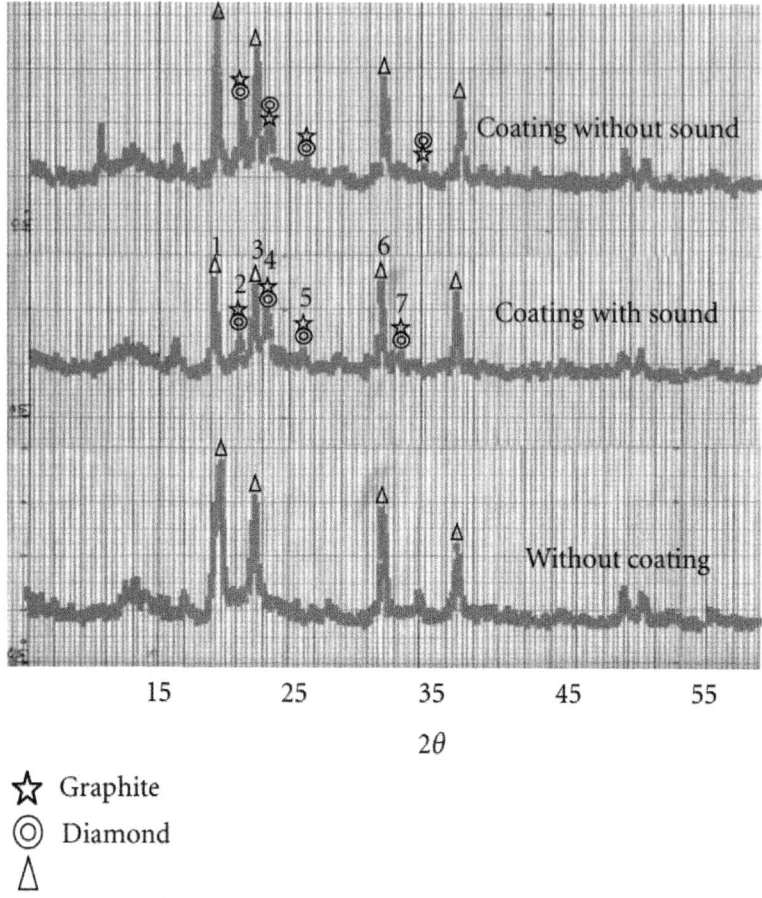

Figure 7: XRD analysis of coated surface without sound vibration at top, with sound vibration at middle, and without coating on stainless steel substrate at bottom (S = stainless steel 304, T_{sub} = 1000°C, T_{act} = 1900C, P_{ch} = 25 torr, and L_{gap} = 5mm).

For comparison of data of area under the significant peaks of Figure 7 indicating the intensity are summarized and presented in Table 4. The peaks 1, 3, and 6 of these figures indicate the presence of austenite of austenitic stainless steel, and the other peaks of the figures indicate graphite and diamond. The intensity of the peaks 1, 3, and 6 of austenite of austenitic stainless steel substrate without coating are 38, 16, and 13 units, while for coated substrate without sound vibration are 30, 17, and 11 units, and for coated substrate with vibration are 17, 12, and 10 units. From these data, it is observed that the intensity of the peaks of austenite of austenitic stainless steel decreases in the coated layer due to deposition of carbon. This intensity of the peaks of austenite of austenitic stainless steel with the coated layer under vibration reduces more as compared with the coated layer under no vibration. On the other hand, the intensity of the peaks 2, 4, 5, and 7 of graphite and diamond under no vibration are 12, 8, 4, and 4 units and for coating with sound vibration are 9, 11, 5, and 5 units. It is seen from these data that the intensity of the peaks of graphite and diamond for the sample coated under vibration increases compared with the coated layer under no vibration that means deposition rate under vibration is better as compared to without vibration.

Table 4: Comparison of intensity among substrate without coating, coating without sound vibration and coating with sound vibration

	Serial number						
	1	2	3	4	5	6	7
	2θ						
Conditions	19.5	21.2	22.5	23.3	26.0	32.6	33.4
	Fe (γ)	Diamond/ gaphite	Fe (γ)	Graphite/ diamond	Diamond/ graphite	Fe (γ)	Graphite/ diamond
	intensity (I)						
Substrate without coating	38		16			13	
Substrate with coating without sound vibration	30	12	17	8	3.5	11	4
Substrate with coating with sound vibration	17	9	12	11	5	10	5

Possible causes of higher deposition rate, compactness and smoothness of the deposited coating under sound vibration condition can be explained as follows.

The complex chemical and physical processes, which occur during diamond CVD, comprise several different but interrelated features. The process gases of the chamber before diffusing toward the substrate surface pass through an activation region (a hot filament), which provide energy to the gaseous species. This activation causes molecules to fragment into reactive radicals and atoms, creates ions and electrons, and heats the gas up to temperatures approaching a few thousand Kelvin. Beyond the activation region, these reactive fragments continue to mix and undergo a complex set of chemical reactions until they strike the substrate surface. At this point, the species is adsorbed and entrapped within the surface, some portions are desorbed again back into the gas phase, or diffuse around close to the surface until an appropriate reaction site is found. If a surface reaction occurs, one possible outcome, if all the conditions are suitable, is diamond. During this process, the addition of sound vibration might increase the energy level of the depositing species. The increase of deposition rate and the surface quality of the deposited coating might be due to elimination or reduction of the potential barrier [14] during the chemical activity by adding some extra sound energy. This extra sound energy may work on the deposition process in the following ways:

- due to the sound media, particles vibrate back and forth and for equilibrium condition, some extra energy remains in the process [15] during the introduction of extra sound vibration into the system;
- at constant temperature, the amount of adsorption depends on pressure [16–18]. Pressure value of sound increases the local pressure;
- as movement of the particles increase, the concentration of diffusing carbon elements increases [19]. Therefore, the diffusion rate of the coating may increase;
- extra vibration of sound may increase the momentum difference of carbon and hydrogen due to their atomic mass difference in methane (CH_4) molecule [15]. This might enhance the chemical reaction in CVD process, and ultimately the deposition rate may increase.

The results obtained under this study shows that sound vibration increases deposition rate with more compact and smoother surface finish. Similar study is conducted to observe the effects of ultrasonic vibrations on the localized electrochemical deposition (LECD) process by Yeo et al. [13]. According to their results, ultrasonic vibrations increase the rate of deposition and improve the concentricity of the fabricated microcolumns. Ultrasonic vibrations perpendicular to the deposition plane improved the uniformity of vapor-deposited films by intensifying diffusion. The effect increases with increasing intensity. Ultrasonic vibrations normal to the deposition surface increase the film thickness. These results confirm our findings in the present study.

CONCLUSIONS

The following can be concluded from this study:

- deposition rate increases significantly (about 18% higher) under sound vibration condition than that of no sound vibration condition;
- the deposition rate under sonic vibration increases slightly with the frequency of vibration;
- the deposition rate under ultrasonic vibration increases significantly with

the frequency of vibration up to a certain value, and after that value, the deposition rate remains almost constant;
- (4) for a particular frequency of vibration, the deposition rate decreases with the distance up to a certain value, and after that the deposition rate remains almost constant (up to observed distance);
- percentage of diamond/graphite in the deposited coating increases about 10% with the addition of sound vibration;
- the surface morphology under SEM analysis of the deposited coatings under sound vibration condition is observed as more compact and smoother surface finish than that of without vibration condition.

Therefore, by maintaining an appropriate level of frequency of vibration and the distance between substrate and activation heater deposition rate of Carbon (diamond/graphite) may be maintained to higher value.

Notations

T_{sub}:	Substrate heater temperature
T_{act}:	Activation heater temperature
L_{gap}:	Distance between substrate and activation heater
P_{ch}:	Pressure of the reactor chamber.

REFERENCES

1. H. O. Pierson, Handbook of Chemical Vapor Deposition, Noyes, Norwich, NY, USA, 2nd edition, 1999.
2. R. F. Bunshah, Handbook of Deposition Technologies for Films and Coatings, Noyes, NJ, USA, 2nd edition, 1994.
3. L. L. Regel and W. R. Wilcox, "Diamond film deposition by chemical vapor transport," Acta Astronautica, vol. 48, no. 2-3, pp. 129–144, 2001.
4. C. H. M. Van Der Werf, H. D. Goldbach, J. Löffler et al., "Silicon nitride at high deposition rate by Hot Wire Chemical Vapor Deposition as passivating and antireflection layer on multicrystalline silicon solar cells," Thin Solid Films, vol. 501, no. 1-2, pp. 51–54, 2006.
5. E. J. Corat and D. G. Goodwin, "Temperature dependence of species concentrations near the substrate during diamond chemical vapor deposition," Journal of Applied Physics, vol. 74, no. 3, pp. 2021–2029, 1993.

6. M. C. McMaster, W. L. Hsu, M. E. Coltrin, D. S. Dandy, and C. Fox, "Dependence of the gas composition in a microwave plasma-assisted diamond chemical vapor deposition reactor on the inlet carbon source: CH_4 versus C_2H_2," Diamond and Related Materials, vol. 4, no. 7, pp. 1000–1008, 1995. ·

7. Y. Fu, C. Q. Sun, H. Du, and B. Yan, "From diamond to crystalline silicon carbonitride: Effect of introduction of nitrogen in CH_4/H_2 gas mixture using MW-PECVD," Surface and Coatings Technology, vol. 160, no. 2-3, pp. 165–172, 2002. · ·

8. L. L. Regel and W. R. Wilcox, "Deposition of diamond on graphite and carbon felt from graphite heated in hydrogen at low pressure," Journal of Materials Science Letters, vol. 19, no. 6, pp. 455–457, 2000. · ·

9. W. Yuan, M. Banan, L. L. Regel, and W. R. Wilcox, "The effect of vertical vibration of the ampoule on the directional solidification of InSbGaSb alloy," Journal of Crystal Growth, vol. 151, no. 3-4, pp. 235–242, 1995. ·

10. L. Chowa, D. Zhoub, A. Hussainb, et al., "Chemical vapor deposition of novel carbon materials," Thin Solid Films, vol. 368, pp. 193–197, 2000.

11. D. S. Dandy and M. E. Coltrin, "Relationship between diamond growth rate and hydrocarbon injector location in direct-current arcjet reactors," Applied Physics Letters, vol. 66, no. 3, 3 pages, 1995. · ·

12. S. Kumar, P. N. Dixit, D. Sarangi, and R. Bhattacharyya, "High rate deposition of diamond like carbon films by very high frequency plasma enhanced chemical vapor deposition at 100 MHz," Journal of Applied Physics, vol. 93, no. 10, pp. 6361–6369, 2003. · ·

13. S. H. Yeo, J. H. Choo, and K. H. A. Sim, "On the effects of ultrasonic vibrations on localized electrochemical deposition," Journal of Micromechanics and Microengineering, vol. 12, no. 3, pp. 271–279, 2002. · ·

14. T. Burakowski and T. Wierzchon, Surface Engineering of Metal, CRC Press, New York, NY, USA, 2000.

15. B. Bhushan, Principles and Applications of Tribology, A Wiley-Interscience, New York, NY, USA, 1999.

16. J. Oscik, Adsorption, E. Horwood Lim, Chichester, UK, 1982.

17. F. C. Tompkins, Chemisorption of Gases on Metals, PWN, Warsaw, Poland, 1985.

18. Joint Report, Physical Chemistry, PWN, Warsaw, Poland, 2nd edition, 1965.
19. S. Mrowec, "Selected topics from the chemistry of defects and theory of diffusion in the solid state," Geological Publication, Warsaw, Poland, 1974.

Chapter 6

A NOISE LEVEL PREDICTION METHOD BASED ON ELECTRO-MECHANICAL FREQUENCY RESPONSE FUNCTION FOR CAPACITORS

Lingyu Zhu[1], Shengchang Ji[1], Qi Shen[2], Yuan Liu[1], Jinyu Li[1], Hao Liu[1]

[1] State Key Laboratory of Electrical Insulation and Power Equipment, Xi'an Jiaotong University, Xi'an, Shaanxi, China
[2] Shaoxing Electric Power Bureau, Shaoxing, Zhejiang, China

ABSTRACT

The capacitors in high-voltage direct-current (HVDC) converter stations radiate a lot of audible noise which can reach higher than 100 dB. The existing noise level prediction methods are not satisfying enough. In this paper, a new noise level prediction method is proposed based on a frequency response function considering both electrical and mechanical characteristics of capacitors. The electro-mechanical frequency response function (EMFRF) is defined as the frequency domain quotient of the vibration response and the squared capacitor voltage, and it is obtained from impulse current experiment. Under given excitations, the vibration response of the capacitor tank is the product of EMFRF and the square of the given capacitor voltage in frequency domain, and the radiated audible noise is calculated by structure acoustic coupling formulas. The noise level under the same excitations is also measured in laboratory, and the results are compared with the prediction. The comparison proves that the noise prediction method is effective.

INTRODUCTION

With the rapid development of high-voltage direct-current (HVDC) transmission, the number of capacitors in HVDC converter stations and the harmonic currents flowing through the capacitors increase dramatically, leading to a great increase of audible noise coming out from the capacitors [1]. The noise may cause serious impact on the life of people around, such as disturbing their peace and endangering their health. Control of the noise has been an important task for researchers and engineers. If the noise level is predicted accurately before a converter station is constructed, corresponding measures against the noise can be taken in advance. Therefore, study of noise level prediction methods is of research value and engineering significance.

Much research has been devoted to studying the characteristics and prediction methods of the capacitor noise. Cox and Guan calculated the vibration responses of capacitor surface under distorted capacitor current based on the transfer functions obtained from impact hammer tests, but the noise level of the capacitors was measured instead of calculated [2]. Smede et al, in their experiment, established a 1:4 scaled acoustic model of capacitor stack to study the characteristics of noise [3]. Obviously, this method is uneconomical, time-consuming and tedious. In our previous paper [4], a formula was presented to calculate the capacitor noise level based on vibration velocity of capacitor surface. However, the calculation of the vibration was not concerned. We have also presented a method for calculating the noise level of capacitors based on modal analysis and impact hammer experiment [5]. This method is currently the most adopted approach by the major manufacturers in China in predicting the noise level when designing converter stations. However, in the impact hammer experiment in[5], the impact force was only applied on the capacitor tank, and the vibration of the capacitor elements under electromagnetic force and the vibration propagation inside capacitor were not taken into account.

In spite of the numerous studies on capacitor noise in HVDC systems, there still lack convincing and satisfying methods for noise level prediction. In this paper, a new noise level calculation approach based on electro-mechanical frequency response function (EMFRF), which is obtained from impulse current experiment, is presented and verified. The paper is organized as follows: In the first section (Noise Prediction Method Based on EMFRF), the definition of EMFRF is given and the noise-level prediction procedure is described. The second section discusses the obtaining of EMFRF from impulse current experiment. In the third section, capacitor noise level is predicted using the presented method and measured experimentally. The prediction results are compared with the experimental data so as to verify the effectiveness of the presented prediction method.

METHODS
Generation of Audible Noise

In general, it is the can-type all-film capacitors that are used in HVDC systems. A capacitor unit is consisted of a steel-covered can and two bushings, the structure of which is shown in Figure 1. The can is filled with oil and contains a capacitor element package formed by a number of capacitor elements connected in series or parallel. The capacitor element is made by winding two aluminum foils and a number of plastic or paper films.

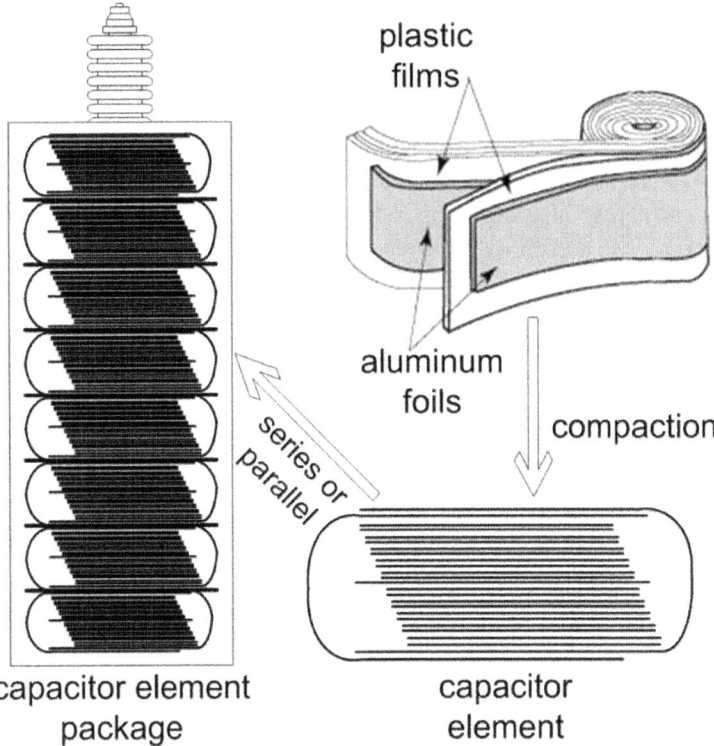

Figure 1. Structure of can-type capacitor.

When voltage is applied on the capacitor, all aluminum foils are energized and nearly all plastic films are in force. Ac capacitor voltages will generate time-varying forces that lead to vibrations. The force in the capacitor element package finally causes vibrations of the steel enclosure of the capacitor unit and thus generates acoustic airborne sound [1], [6].

Definition of EMFRF

A capacitor is assumed to be a linear mechanical system, which can be described by frequency response function $H_M(w)$ as follows:

$$H_M(\omega) = \frac{V(\omega)}{F(\omega)} \tag{1}$$

where ω is the vibration angular frequency, V(w) is the vibration velocity response of the capacitor tank in frequency domain, and F(w) is the attractive electric force in frequency domain. As analyzed in [5], the attractive electric force f(t), which is the motivation of capacitor vibration, is proportional to the square of the voltage applied on the capacitor. This relation is expressed as

$$f(t) = Ku^2(t) \tag{2}$$

where K is the proportional coefficient, and u(t) is the ac voltage applied on the capacitor. As a result, the system composed of the vibration response of the capacitor tank and the squared voltage applied on the capacitor is also linear. A frequency response function containing both electrical and mechanical characteristics is therefore employed to describe the system:

$$H_{EM}(\omega) = \frac{V(\omega)}{F\{u^2(t)\}} \tag{3}$$

where $F\{u^2(t)\}$ is the spectrum of the square of the voltage. The function $H_{EM}(w)$ is named electro-mechanical frequency response function (EMFRF).

Noise Level Prediction Method Based on EMFRF

Firstly, the voltage applied on a capacitor is calculated from the capacitance of the capacitor and the current flowing through it. The spectrum of the square of the voltage can be obtained by using Fourier transform method. This step has been described in [1].

Secondly, EMFRF of the capacitor, $H_{EM}(w)$, is obtained by impulse current experiment, which will be described in the section of "Obtaining of EMFRF from Impulse Current Experiment".

Then the corresponding vibration velocity response of the capacitor tank is calculated by

$$V(\omega) = H_{EM}(\omega) F\left[u^2(t)\right] \tag{4}$$

Finally, the audible noise level can be predicted from the vibration velocity. Suppose that the vibration velocity of a capacitor surface is v, and then its radiated sound power W_{rad} is

$$W_{rad} = \rho c S v^2 \sigma \tag{5}$$

where r is the air density, c is the sound velocity in air, S is the area of the surface radiating sound, v is the vibration velocity, and σ is the radiation ratio (no unit) [1]. The size of vibration source is much smaller than the main vibration wavelength. Therefore, according to [7], can be estimated by (6), which is demonstrated in Appendix S1.

$$\log \sigma = -\log\left[1+0.1\frac{c^2}{(f_v d)^2}\right] \qquad (6)$$

where d is the feature size of sound source, $d \approx \sqrt{S/\pi}$, and f_v is the vibration frequency.

Meanwhile, the sound power level L_W in the unit of decibels (dB) is defined by

$$L_W = 10\log\frac{W_{rad}}{W_0} \qquad (7)$$

where W_0 is the sound reference power of 10^{-12} W [1].

Substitute (5) into (7), we have

$$L_W = 10\log\frac{v^2}{v_0^2} + 10\log\frac{S}{S_0} + 10\log\frac{\sigma}{\sigma_0} + 10\log\frac{\rho c}{(\rho c)_0} \qquad (8)$$

where v_0 is the reference velocity of 5×10^{-8} m/s, S_0 is the reference area of 1 m2, s_0 is the reference radiation ratio of 1, and $(r_c)_0$ is the reference air sound impedance of 400 kg/(m²s).

In a semi-anechoic room (a room where sound reflections only come from the floor because the walls and ceiling are absorbent), the relation between sound power level L_W and sound pressure level L_P is

$$L_P = L_W - 10\log(2\pi r^2) \qquad (9)$$

where r is the distance from the measurement point to the sound source in the unit of m [1].

Generally, A-weighted sound pressure level is widely used to evaluate the strength of noise. The frequency characteristics of A-weighting network are shown in Figure 2. Measurement adopting "A-weighting" is in the unit of dB(A), and generally agrees with people's assessment of "loudness". The A-weighted sound pressure level L_{PA} is calculated from the sound pressure level at each frequency by

$$L_{PA} = 10\log\left[\sum 10^{(L_{P_i}+\Delta A_i)/10}\right] \qquad (10)$$

where L_{P_i} is the sound pressure level (in dB) at the ith frequency band, DA_i is the A-weighting gain (in dB) at the ith frequency band [1].

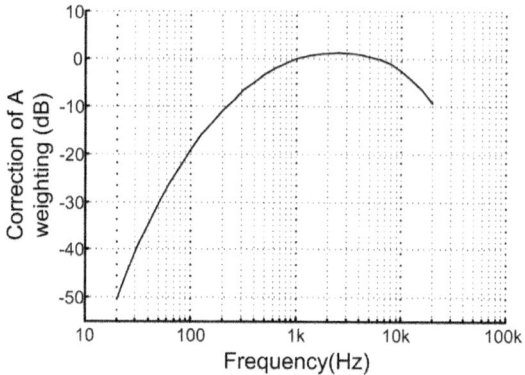

Figure 2. Frequency characteristics of A-weighting network.

The flowchart of audible noise level prediction based on EMFRF is shown in Figure 3.

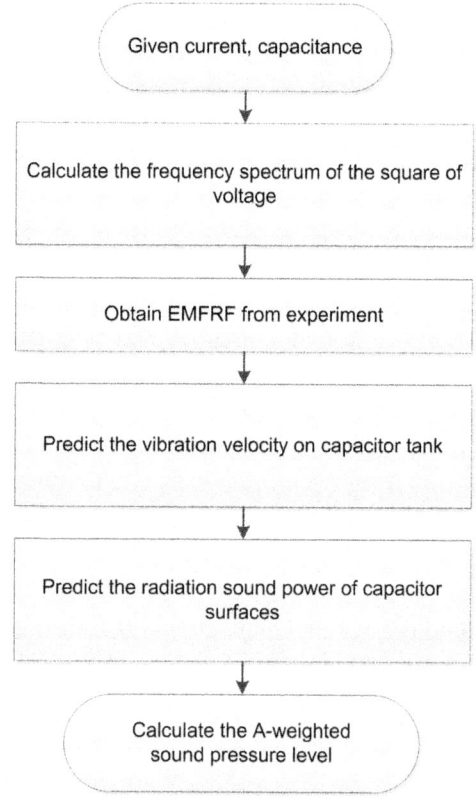

Figure 3. Flowchart of the noise level prediction based on EMFRF.

Experimental System for Measuring EMFRF

In order to learn the electro-mechanical characteristics of a capacitor in wide frequency band, impulse current experiment is employed to obtain EMFRF.

The experimental system for measuring EMFRF is shown in Figure 4. The experiment procedure is similar to that of the short-circuit discharge test in type tests of capacitors [8]. The capacitor is charged to a voltage U_0 by the half-wave rectifier consisted of test transformer and high voltage silicon stack. The sphere gap is triggered and the capacitor discharges through a small resistor. Simultaneously, the vibration velocity of capacitor tank is measured by a portable digital vibrometer (PDV), and the impulse current applied to the capacitor is measured by the shunt. Both the impulse current signal and the vibration velocity signal are acquired by the oscilloscope.

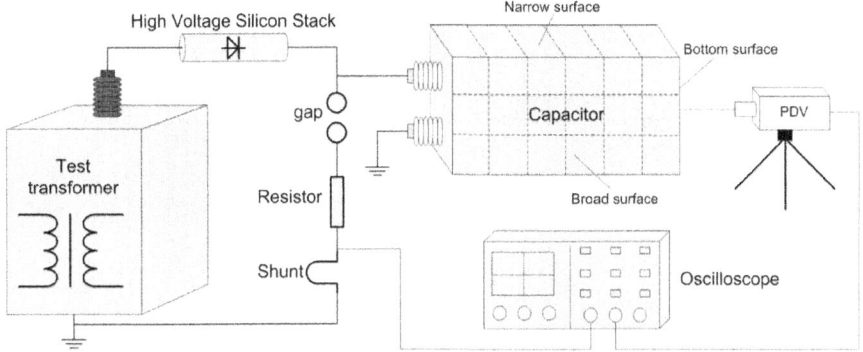

Figure 4. Experimental system for measuring EMFRF.

The parameters of the devices are as follows:
- **Shunt:** Resistance of 0.00184 W,
- **PDV:** PDV-100 vibrometer manufactured by Polytec GmbH, working frequency range of 0-22 kHz, propagation delay of approximately 1 ms, measurement range of 20 mm/s, 100 mm/s or 500 mm/s (adjustable via display), low pass filter cutoff frequency (0.1 dB) of 1 kHz, 5 kHz or 22 kHz (adjustable via the display),
- **Oscilloscope:** DPO4054 oscilloscope manufactured by Tektronix, bandwidth of 500 MHz; sampling rate of 5 GS/s, record length of 20 M,
- **Capacitor:** Rated voltage of 12 kV, rated capacity of 417 kVar, measured capacitance of 9.46 mF.

Configuration and dimension of the capacitor are shown in Figure 5. The surface where the bushings are installed is appointed as top surface, and the opposite is the bottom surface. The other four surfaces are called narrow surface or broad surface according to their breadth.

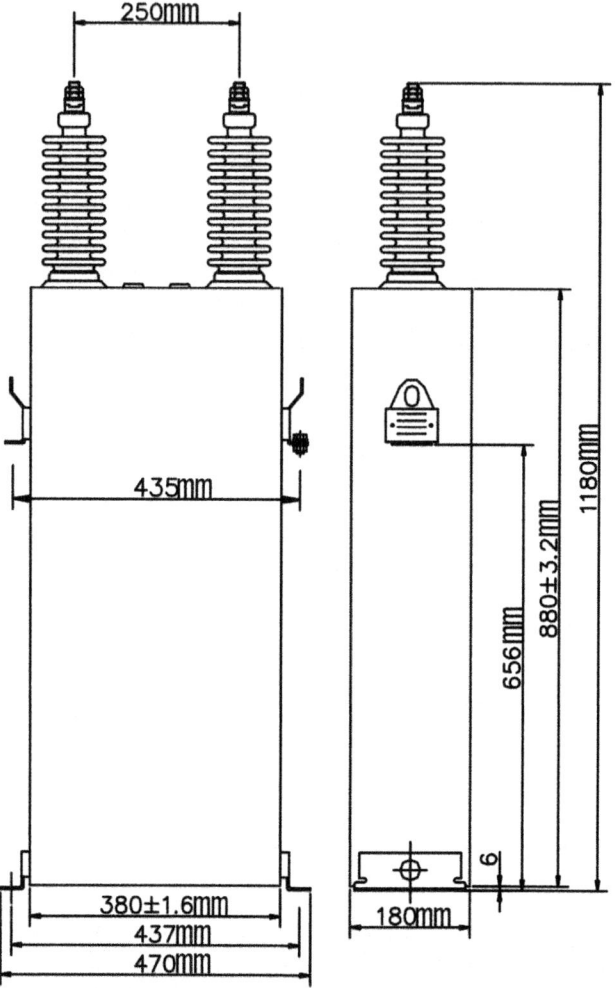

Figure 5. Configuration and dimension of the capacitor.

The capacitor is fixed on a steel frame with the narrow side upward as in the field (Figure 6).

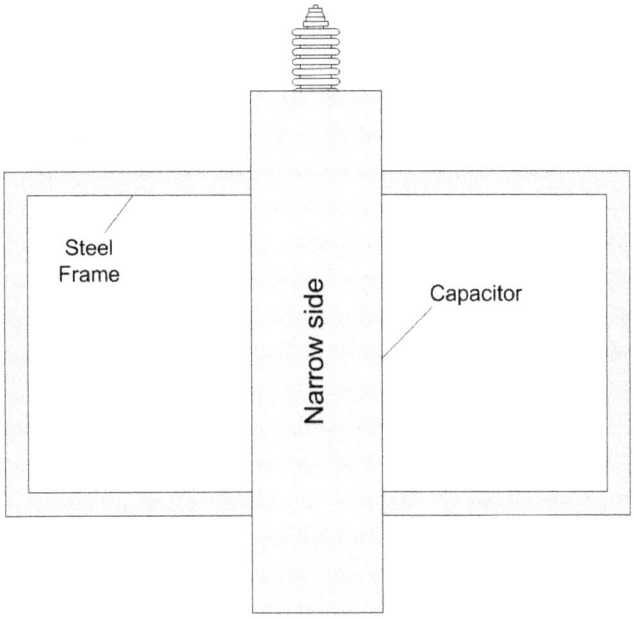

Figure 6. Fixing of the capacitor (top view).

Calculation of the Spectrum of the Squared Voltage

The small resistor is adjusted in advance to make sure that the impulse current contains abundant frequency components. The waveform of the impulse current is shown in Figure 7. The voltage on the capacitor is the integral of the current, which is

$$u(t) = \frac{1}{C} \int_{\tau=0}^{t} i(\tau) d\tau + U_0 \tag{11}$$

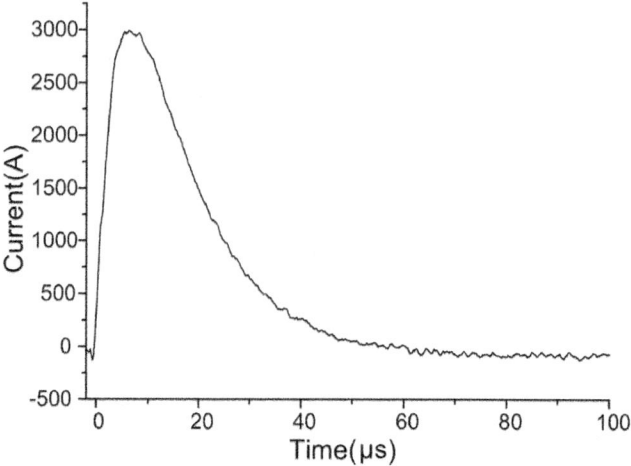

Figure 7. Waveform of the impulse current.

As shown in Figure 8, the waveform of the voltage has a steep trailing edge when the impulse current flows through the capacitor. Obviously, the square of the voltage is infinite and aperiodic in time domain, so its spectrum cannot be calculated directly by Fourier transform. However, the derivative of the squared voltage is finite and the spectrum can be calculated. The spectrum of the squared voltage can be then obtained based on the derivative property of Fourier transform.

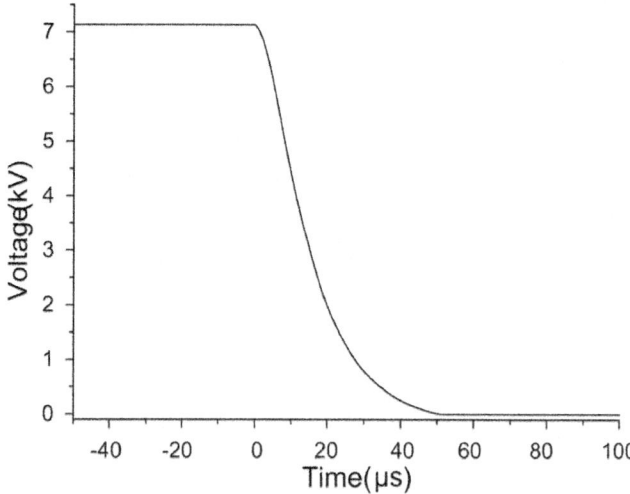

Figure 8. Waveform of capacitor voltage.

The derivative of the square of voltage is

$$\frac{d(u^2(t))}{dt} = 2u(t)\frac{du(t)}{dt} \tag{12}$$

The capacitor current is the derivative of the voltage, which is

$$i(t) = C\frac{du(t)}{dt} \tag{13}$$

Synthesize (12) and (13), we can get

$$\frac{d(u^2(t))}{dt} = \frac{2}{C}u(t)i(t) \tag{14}$$

According to the derivative property of Fourier transform, if $x(t) \leftrightarrow X(\omega)$, there is $\frac{dx(t)}{dt} \leftrightarrow j\omega X(\omega)$. As a result, the Fourier transform of u^2 can be calculated as

$$F[u^2(t)] = \frac{2}{j\omega C}F[ui] \tag{15}$$

Calculation and Analysis of EMFRF

A number of n equally spaced points are marked over the surfaces of the capacitor tank. The impulse current is applied to the capacitor repeatedly. Simultaneously, the vibration velocities of the marked points are recorded sequentially. The vibrations at each point are measured at least two times to avoid the influence of random error on the measurement. In fact, at all the measurement points, the two vibrations measured at different time are exactly the same as shown in Figure9, which means capacitor responses identically under same electric excitations and the reproducibility of the calculation method is satisfying.

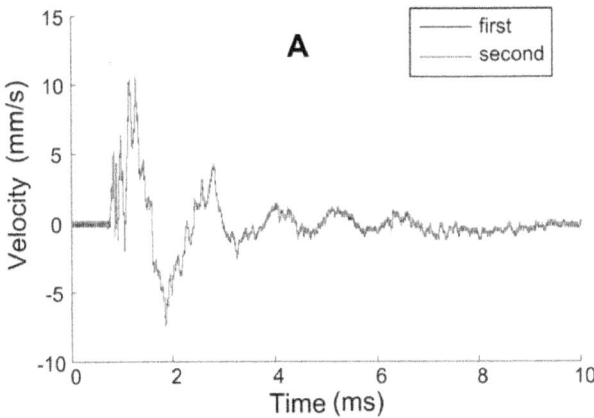

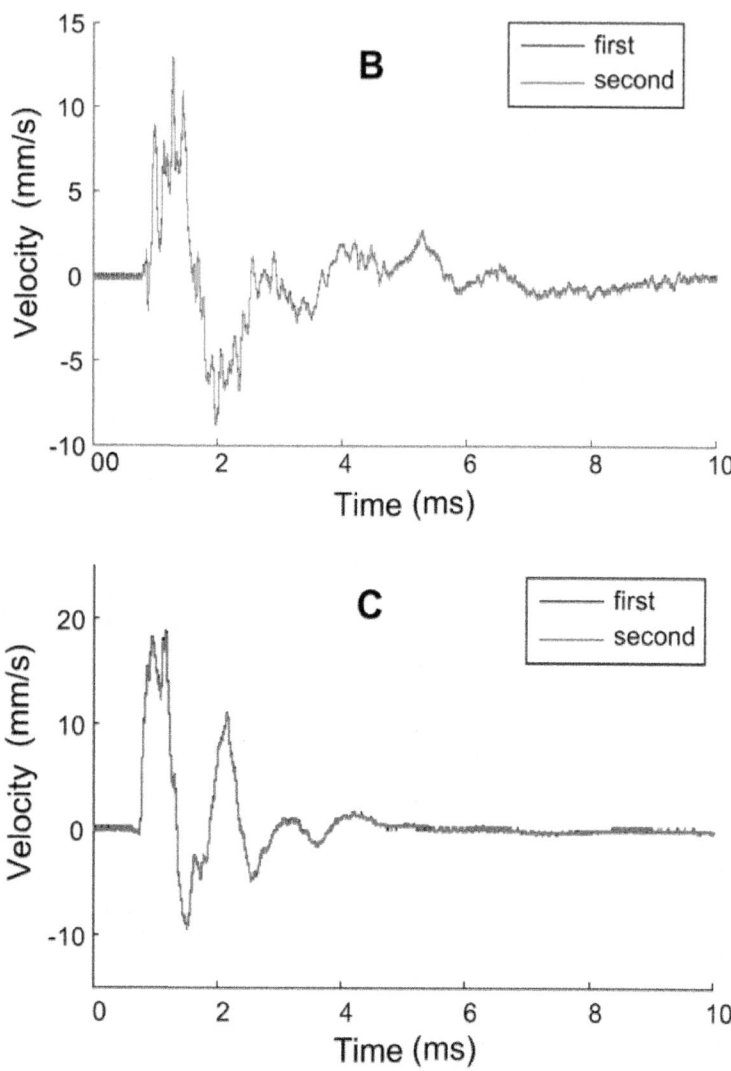

Figure 9. Two vibrations measured at different time. A Broad surface, B Narrow surface, and C Bottom surface.

Suppose that when the impulse current i_j is applied, the recorded vibration velocity at the jth point is $v_j(t)$ and the voltage is $u_j(t)$. Synthesize (3) and (15), we have

$$H_j(\omega) = \frac{j\omega CF[v_j(t)]}{2F[u_j(t)i_j(t)]} \qquad (16)$$

where j=1,2,...,n.

The magnitude plots of EMFRF for the broad, narrow and bottom surfaces of the tested capacitor are shown in Figure 10. The plots indicate that the vibration responses of side surfaces are quite similar, and that of bottom surface is much different. The magnitude of the side surface EMFRF is very small for low frequency (under 3 kHz). When the frequency is above 3 kHz, there are a large number of resonant frequencies, represented by the peaks of EMFRF magnitude. However, the magnitude of EMFRF of bottom surface is big under 3 kHz and small above 3 kHz. The peak value of bottom surface is about 4-6 times that of side surfaces, which is consistent with the conclusion that the bottom surface is the main source of noise radiation [1].

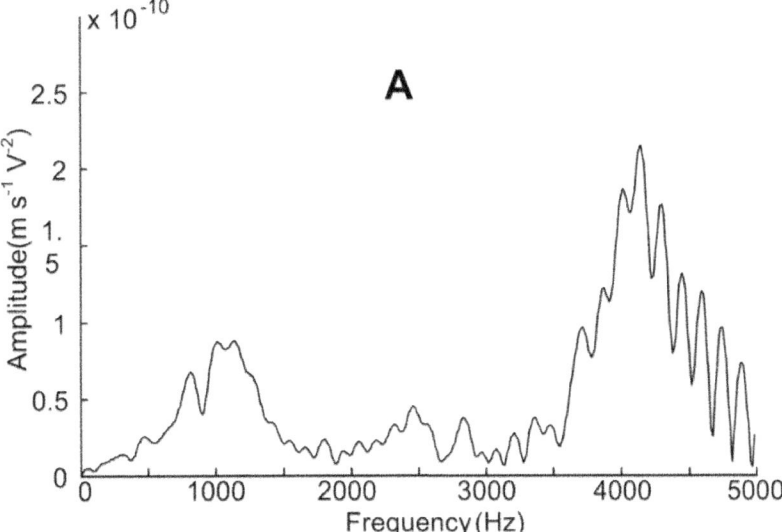

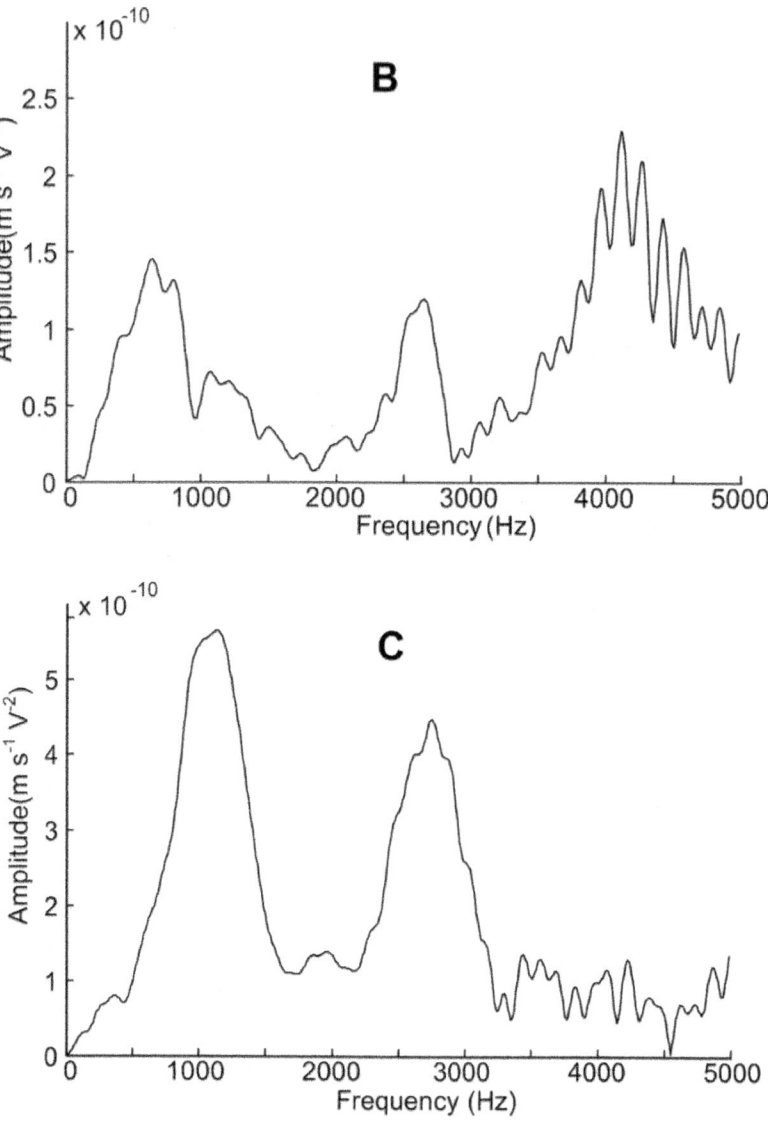

Figure 10. EMFRF of capacitor surfaces. A Broad surface, B Narrow surface, and C Bottom surface.

In this experiment, the impulse current flows through the capacitor and the capacitor elements are motivated to vibrate, so that the obtained EMFRF contains the information of vibration generation and propagation in addition to the mechanical features of the capacitor. Compared with the calculation method presented in [5], the method proposed here has a better accuracy. In [5], frequency response functions which describe the natural vibration characteristics of capacitor tank were obtained from impact hammer experiment. The impact force was only applied on the surface of capacitor tank in that experiment, so the vibration of the capacitor elements under electromagnetic force and the vibration propagation inside capacitor were not taken into account.

Laboratory Measurement of Capacitor Noise

The noise levels of the capacitor under certain operating conditions are measured in laboratory. The experimental data are compared with the calculation results of the presented prediction method to verify the effectiveness of the method.

The noise measuring system, shown in Figure 11, consists of current measuring part and noise measuring part. The main circuit is made up of the reactor, the capacitor and the harmonic current source. The reactor and the capacitor form the current resonance circuit. The frequency of the harmonic current source can be controlled within the range of 0-600 Hz. A sound pressure spectrum analyzer, marked as SPL in Figure 11, is used to record the audible noise data. The capacitor and the analyzer are placed in a semi-anechoic room to avoid interference from environmental noise.

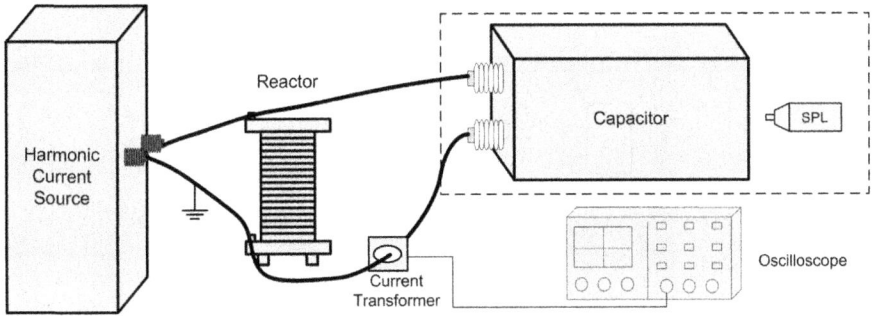

Figure 11. Noise measuring system.

The parameters of the devices are as follows:
- **Sound pressure spectrum analyzer:** NA-28 sound-level meter and the 1/3 octave band real-time analyzer manufactured by RION in Japan, sensitivity of -27 dB and measurement range from -25 dB to 130 dB,
- **Semi-anechoic room:** Space size of 6.3 m5.5 m5.4 m, frequency range from 100 Hz to 10 kHz for free field, background noise of 23.9 dB.
- **Oscilloscope:** DPO4054B oscilloscope manufactured by Tektronix, bandwidth of 500 MHz, sampling rate of 5 GS/s, record length of 20 M.

The sound pressure level of the filter capacitor is measured under five different currents. The frequencies and amplitudes of the test currents are shown in Table 1. To avoid the impact of the mechanical fixing on the radiating noise level, the capacitor is fixed in the same way as in the experiment for measuring EMFRF.

Table 1. Test current

Frequency/Hz	150	250	350	450	550
Current/A_{rms}	34.75	34.75	34.75	34.75	34.75

doi:10.1371/journal.pone.0081651.t001

Measurement points for the audible noise are arranged at 30 cm and 50 cm away from the center of each surface, as shown in Figure 12. Compared with the arrangement in [9], the arrangement in this paper adds the measurement points at 30 cm to verify the effectiveness of the presented prediction method under different distances. The data of current flowing through the capacitor is acquired simultaneously with audible noise at the measurement points.

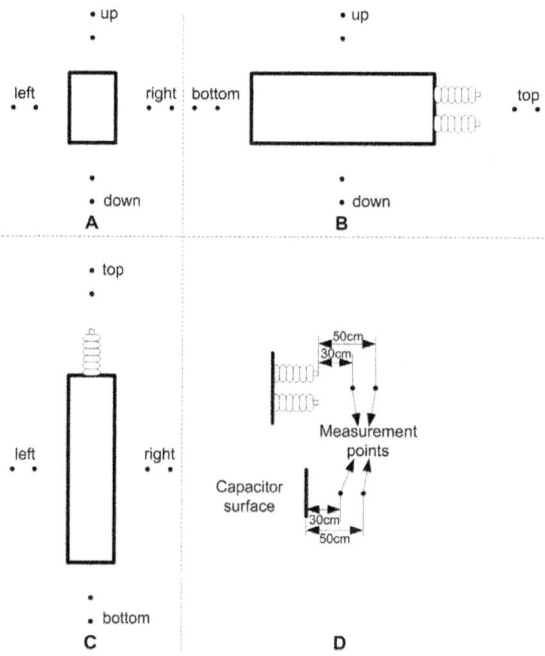

Figure 12. Measurement points for audible noise. A Front view, B Right View, C Top view, and D measurement-point arrangement.

Calculation of Sound Pressure Level via the Presented Prediction Method

At a given current angular frequency w and current amplitude I, the voltage of the capacitor is a single frequency signal with the angular frequency of w. Its amplitude can be calculated by

$$U = \frac{I}{j\omega C} \quad (17)$$

The square of the voltage can be expressed as

$$u^2 = U^2 \sin^2(\omega t + \varphi) = \frac{U^2}{2} - \frac{U^2}{2}\cos(2\omega t + 2\varphi) \quad (18)$$

It can be seen that the square of the voltage contains a dc component and a sinusoidal component with the angular frequency of 2w and the amplitude of $U^2/2$.

As expressed in (4), the vibration responses of each surface equal the frequency domain product of the squared voltage and the calculated EMFRF. The vibration velocity at each of the marked points for calculating the noise level can be predicted. The geometric average of the predicted velocities at all the points over each surface is adopted as the vibration velocity of the corresponding surface:

$$v_s = \frac{1}{N}\sqrt{\sum_{i=1}^{N} v_i^2} \quad (19)$$

where n_s is the surface vibration velocity, N is the number of measurement points, n_i is the predicted velocity at the ith point. The calculation results are shown in Table 2.

Table 2. Calculated vibration velocity

Current frequency/Hz	Voltage/kV	Vibration frequency/Hz	Vibration velocity/mm/s		
			bottom surface	broad surface	narrow surface
150	3.9	300	0.59	0.15	0.17
250	2.3	500	0.29	0.11	0.12
350	1.7	700	0.21	0.12	0.08
450	1.3	900	0.32	0.08	0.03
550	1.1	1100	0.25	0.04	0.02

doi:10.1371/journal.pone.0081651.t002

Based on the calculated vibration response, the sound pressure level of each surface can be predicted by (8) to (10).

RESULTS AND DISCUSSION

Comparison of Measurement and Prediction

The measurement results and prediction results are compared, as shown in Table 3 and Table 4. The predicted noise level is close to the measured data at most points, suggesting that the presented method can effectively predict the noise level of capacitors in HVDC systems. There exist certain differences at a couple of points under certain excitations, for example, the 30 cm point of bottom surface under 350 Hz excitation and the 50 cm point of narrow surface points under 450 Hz excitation. The discrepancies are mainly caused by the simplifications in the model. The geometric average velocity of all the points on each surface is taken as the vibration velocity of the whole surface, which may cause errors in calculation.

Table 3. Comparison between the Measured and the Predicted Noise Levels at 30(dB(A))

Frequency of current/Hz	Surface	Measurement	Prediction	Deviation
150	Bottom	56.4	56.9	0.9%
	Broad	50.8	53.3	4.9%
	Narrow	51.6	51.1	1.0%
250	Bottom	61.4	59.7	2.8%
	Broad	56.2	57.4	2.1%
	Narrow	56.0	56.0	0.0%
350	Bottom	67.3	63.6	5.5%
	Broad	59.2	60.4	2.0%
	Narrow	58.1	55.7	4.1%
450	Bottom	69.9	68.4	2.1%
	Broad	63.0	61.1	3.0%
	Narrow	54.8	53.2	2.9%
550	Bottom	65.1	65.3	0.3%
	Broad	56.8	55.9	1.6%
	Narrow	46.8	46.0	1.7%

doi:10.1371/journal.pone.0081651.t003

In addition, it is assumed that the capacitor in our study is a linear mechanical system, but it is not perfectly linear actually. What's more, equation (9) is valid for far field conditions, but the measurements and the calculations in this study are conducted for the near field. The hemispherical sound propagation mode is chosen to simplify the model [1]. Comparison of the calculation results and the measurement results indicates that the simplification is acceptable, but it still may bring errors into calculation. Lastly, the capacitor current in the noise measuring experiment may also vary with the power grid fluctuation, making the measurement results deviate from the real values.

Table 4. Comparison between the Measured and the Predicted Noise Levels at 50(dB(A))

Frequency of current/Hz	Surface	Measurement	Prediction	Deviation
	Bottom	52.5	52.4	0.2%
150	Broad	47.2	49.4	4.7%
	Narrow	46.6	48.3	3.6%
	Bottom	56.1	56.3	0.4%
250	Broad	50.9	53.4	4.9%
	Narrow	54.0	52.50	2.8%
	Bottom	61.9	60.2	2.7%
350	Broad	55.4	56.5	3.8%
	Narrow	55.2	53.7	2.7%
	Bottom	64.0	63.0	1.6%
450	Broad	59.2	56.9	3.9%
	Narrow	52.7	49.2	6.6%
	Bottom	62.1	61.9	0.3%
550	Broad	56.3	54.6	3.0%
	Narrow	45.6	44.1	3.3%

doi:10.1371/journal.pone.0081651.t004

Limitations of the Study, Open Questions and Future Work

In this paper, the deviations in the comparison are just qualitatively analyzed. Some possible factors causing the discrepancies are given. In the future, we plan to quantify the errors caused by these factors and amend the calculation method accordingly. The nonlinearity of the system will be modeled and calculated to evaluate the nonlinear error in the calculation. The error brought by the substitution of the geometric average velocity for the actually velocity will be analyzed by the simulation software Sysnoise. In Sysnoise, the two modes for velocity boundary should be both employed, and the discrepancy in the calculation results will show the quantitative error. At last, the noise will be measured continuously in the future research and stable results will be chosen to lower the measurement error.

The predicted noise should be compared with the measured result at a distance much further than 50 cm since the noise level in the nearby residential district is the greatest concern. However; the noise measurement distance is restricted by the size of the semi-anechoic room in this paper. According to ISO 3745-2003 [10], the noise measurement is conducted in a semi-anechoic room to provide a free field over a reflecting plane without environment interference. The size of the semi-anechoic room is 6.3 m5.5 m5.4 m. As a result, the measurement distance cannot be any further. Nevertheless, we believe that the comparisons at the two distances are convincing enough to prove the effectiveness of the proposed method. In the further research, we plan to compare the predicted noise with the sound field simulation result at a further distance to refine our work.

Only the noise prediction method for single capacitors is proposed and verified in this paper, the calculation method for multiple capacitors is not involved. To do this, it should be firstly proved that the capacitors of a same type radiate same noise under same excitations. In [5], the uniformity has been assumed because the inside structure is uniform, which is theoretically reasonable. Besides, the discrepancy between the noises radiated from capacitors of a same type can also be investigated by experiment to prove the uniformity, which will be our future work. Another trouble in the calculation of multiple-capacitors noise is the superimposition of the noise radiated from independent sources. The sound field becomes complicate in the case of multiple sources, so it still needs further study.

CONCLUSIONS

In this paper, a novel noise level prediction method is presented, which is based on a frequency response function considering both electrical and mechanical characteristics of capacitors. The "electro-mechanical frequency response function (EMFRF)" is defined as the frequency domain quotient of the vibration response and the squared voltage. EMFRF can be obtained from impulse current experiment. The vibrations of the capacitor surfaces under given excitations can be predicted based on the measured EMFRF, and the noise levels can then be calculated. The prediction results are compared with the measured data in a laboratory experiment. In spite of some differences, the predicted noise levels are close to the measured values at most measurement points. This indicates that the noise level prediction method for capacitors in HVDC converter stations presented in this paper is effective.

The impulse current experiment in this study uses electric excitations, which is the same as the practical situation. Therefore, the EMFRF obtained from the experiment can reflect not only the natural mechanical characteristics but also the electrical characteristics of capacitors. EMFRF builds the connection between electrical excitations and vibration responses of capacitor tanks, which is not achieved in other methods. This is a great progress in noise level prediction for capacitors in HVDC converter stations.

SUPPORTING INFORMATION

Appendix S1. Demonstration of calculation formula for radiation ratio

If the size of vibration source is much smaller than the main vibration wavelength, the vibration source can be regarded as spherical source. For a spherical sound source[10],

$$\sigma = \frac{k^2 a^2}{1+k^2 a^2} \quad (1)$$

where k is the wavenumber, $k = 2\pi f_v / c$; a is the radius of the spherical source, $k2 = d$. Substitute k and a into (1), we get

$$\sigma = \left[1 + \frac{c^2}{\pi^2 (f_v d)^2}\right]^{-1}$$

Therefore

$$\log \sigma \approx -\log\left[1 + 0.1\frac{c^2}{(f_v d)^2}\right]$$

AUTHOR CONTRIBUTIONS

Conceived and designed the experiments: LZ SJ. Performed the experiments: LZ QS YL JL. Analyzed the data: LZ HL. Wrote the paper: LZ SJ.

REFERENCES

1. CIGRE No.202 W.G 14.26 (2002) HVDC stations audible noise.

2. Cox MD, Guan HH (1994) Vibration and audible noise of capacitors subjected to nonsinusoidal wave-forms. IEEE Transactions on Power Delivery 9: 856–862. doi: 10.1109/61.296267
3. Smede H, Johansson C, Winroth O, Schutt H (1995) Design of HVDC converter stations with respect to audible noise requirements. IEEE Transactions on Power Delivery 10: 747–758. doi: 10.1109/61.400856
4. Cao T, Ji S, Wu P, Zhang J, Li Y, et al. (2010) Calculation method of noise level for capacitor based on vibration signal. Transactions of China Electrotechnical Society 25: 172–177.
5. Ji S, Wu P, Qiaogen Z, Yanming L (2010) Study on the noise-level calculation method for capacitor stacks in HVDC converter station. IEEE Transactions on Power Delivery 25: 1866–1873. doi: 10.1109/tpwrd.2010.2047409
6. William TT, Dahleh MD (1998) Theory of Vibration with Applications (Fifth Edition). Prentice-Hall.
7. GB/T 16539-1996 (1996) Acoustic-Determination of sound power levels of noise sources using vibration velocity- Measurement for seal machinery.
8. IEC 60871-1-1997 (1997) Shunt capacitors for ac power systems having a rated voltage above 1000v- Part 1: General- Performance, testing and rating c safety requirements c guide for installation and operation.
9. Wu P, Ji S, Li Y, Cao T (2011) Study on an audible noise reduction measure for filter capacitors based on compressible space absorber. IEEE Transactions on Power Delivery 26: 438–445. doi: 10.1109/tpwrd.2010.2070880
10. ISO 3745 -2003 (2003) Acoustics determination of sound power levels of noise sources using sound pressure Precision methods for anechoic and hemi-anechoic rooms.

Chapter 7

EXPERIMENTAL INVESTIGATIONS OF NOISE CONTROL IN PLANETARY GEAR SET BY PHASING

S. H. Gawande[1] and S. N. Shaikh[2]

[1]Department of Mechanical Engineering, M. E. Society's College of Engineering, Pune, Maharashtra, India

[2]Department of Mechanical Engineering, AISSMS College of Engineering, Pune, Maharashtra, India

ABSTRACT

Now a days reduction of gear noise and resulting vibrations has received much attention of the researchers. The internal excitation caused by the variation in tooth mesh stiffness is a key factor in causing vibration. Therefore to reduce gear noise and vibrations several techniques have been proposed in recent years. In this research the experimental work is carried out to study the effect of planet phasing on noise and subsequent resulting vibrations of Nylon-6 planetary gear drive. For this purpose experimental set-up was built and trials were conducted for two different arrangements (i.e., with phasing and without phasing) and it is observed that the noise level and resulting vibrations were reduced by planet phasing arrangement. So from the experimental results it is observed that by applying the meshing phase difference one can reduce planetary gear set noise and vibrations.

INTRODUCTION

Gears are essential parts of many precision power and torque transmitting machine such as an automobile. The major functions of a gearbox are to transform speed and torque in a given ratio and to change the axis of rotation. Planetary gears yield several advantages over conventional parallel shaft gear systems. They produce high speed reductions in compact spaces, greater load sharing, higher torque to weight ratio, diminished bearing loads, and reduced noise and vibration. They are used in automobiles, helicopters, aircraft engines, heavy machinery, and a variety of other applications. Despite their advantages, the noise induced by the vibration of planetary gear systems remains a key concern. Planetary gears have received considerably less research attention than single mesh gear pairs. There is a particular scarcity of analysis of two planetary gear systems and their dynamic response. This paper focus on the study of two PGTs with different phasing (angular positions) while keeping every individual set unchanged.

Planetary gear systems are used to perform speed reduction due to several advantages over conventional parallel shaft gear systems. Planetary gears are also used to obtain high power density, large reduction in small volume, pure torsional reactions, and multiple shafting. Another advantage of the planetary gearbox arrangement is load distribution. Because the load being transmitted is shared between multiple planets, torque capability is greatly increased. The more the planets in the system, the greater the load ability, and the higher the torque density. The planetary gearbox arrangement also creates greater stability due to the even distribution of mass and increased rotational stiffness. Despite their advantages the noise induced by vibrations of planetary is concern, particularly in automotive industry where the vehicle interior noise is a key quality metric.

Extensive research work has been carried out by many researchers on the analysis of errors, dynamic response, and noise and vibration reduction in single planetary gears. Schlegel and Mard [1] proposed one strategy of reducing planet gear vibrations which used planet phasing, where the planet configuration and tooth numbers are chosen such that self-equilibration of the mesh forces reduce the net forces and torques on the sun, ring, and carrier to reduce vibration and noise up to 11 dB. Seager [2] explained a more detailed analysis using a static transmission error model of the dynamic excitation. Palmer and Fuehrer [3] also demonstrated the effectiveness of planet phasing and support their arguments with limited experiments. Kaharamam and Blankership [4] studied the use of planet phasing in the context of helical planetary systems in which author used the static transmission error to represent the dynamic excitation in a lumped parameter dynamic model. All of these studies focus on planetary gears with equally spaced planet gears. Parker [5] provided physical explanation for the effectiveness of planet phasing to suppress planetary gear vibration based on the physical forces acting at the sun, planet, and ring meshes. He demonstrated that phenomenon with a dynamic finite element/contact mechanics simulation. Gill-Jeong [6] analyzed a new method of reducing vibrations of spur gear. He did numerical study on reducing the vibration of spur gear pairs with phasing. This new method is based on reducing the variation in mesh stiffness by adding another pair of gears with half-pitch phasing. This reduces the variation in the mesh stiffness of the final (phasing) gear, because each gear compensates for the variation in the other's mesh stiffness. Chen and Ishibashi [7] have investigated the relationship between the meshing phase difference and torsional vibration of planet gears from the standpoint of their rotational meshing cycle. Using planetary gear sets with and without a meshing phase difference, measurements were made of their noise and vibration acceleration under various driving conditions. The method of finishing the gears, the tooth profile contact ratio, and other factors were varied in order to compare and analyze the measured data. In present work method proposed in [7] is extended to reduce the noise level and resulting vibrations of planetary gear set. Meshing of the gears in the planetary gear set that forms the ratio-changing mechanism of an automatic transmission produces gear noise over a wide range of driving conditions from low to high vehicle speed. As per the literature survey it is observed that the internal excitation caused by the variation in tooth mesh stiffness is a key factor in causing vibrations [8, 9]. In order to study the phenomenon of noise control in planetary gear set it is necessary to study the dynamic behavior of gear train.

DYNAMIC MODELS OF GEAR TRAIN

To understand and control gear noise, it is necessary to have knowledge not only about the gears, but also about the dynamic behavior of the system consisting of gears, shafts, bearings, and gear train casing. While designing the gear train the noise characteristics of a gear train can be controlled at the drawing board, because all the components have an important effect on the acoustical output [10]. For relatively simple gear systems it is possible to use lumped parameter dynamic models with springs, masses, and viscous damping. For more complex models, which include, for example, the gear train casing, finite element modeling and analysis is often used. The first dynamic models were used to determine dynamic loads on gear teeth and were developed in the 1920s; the first mass-spring models were introduced in the 1950s [11].

Lumped Parameter Dynamic Models

Özgüven and Houser [11] reviewed the literature on mathematical models used in gear dynamics from 1915 and up to 1986. In this work extensive literature review is carried out. They classified the models in five groups;

- Simple dynamic factor models: this group includes most of the early studies in which a dynamic factor that can be used in gear root stress formulae is determined. These studies include empirical and semiempirical approaches as well as recent dynamic models constructed just for the determination of a dynamic factor.

- Models with tooth compliance: there are a very large number of studies that include only the tooth stiffness as the potential energy storing element in the system. That is, the flexibility of shafts, bearings, and so forth is all neglected. In such studies the system is usually modeled as a single-degree-of-freedom spring-mass system. There is an overlap between the first group and this group since such simple models is sometimes developed for the sole purpose of determining the dynamic factor.
- Models for gear dynamics: such models include the flexibility of the other elements as well as the tooth compliance. Of particular interest have been the torsional flexibility of shafts and the lateral flexibility of the bearings and shafts along the line of action.
- Models for geared rotor dynamics: in some studies, the transverse vibrations of a gear carrying shaft are considered in two mutually perpendicular directions, thus allowing the shaft to whirl. In such models, the torsional vibration of the system is usually considered.
- Models for torsional vibrations: the models in the third and fourth groups consider the flexibility of gear teeth including a constant or time varying mesh stiffness in the model. However there is also a group of studies in which the flexibility of gear teeth is neglected and a torsional model of a geared system is constructed by using torsionally flexible shafts connected with rigid gears.

The studies in this group may be viewed as pure torsional vibration problems, rather than gear dynamic problems. In a study by Cheng [12], the vibrations of spur gears were simulated. The dynamics of the gears was modelled as a nonlinear time-correlated, stationary stochastic process. As excitation of the system, random and harmonic transmission error was used. The vibrations excited by random and harmonic transmission error and time varying mesh stiffness were investigated at different speeds and different loads. Optimization of gear parameters were made to avoid resonance. Kahraman and Singh [13] used a two-degree-of-freedom model of a spur gear pair with backlash, to investigate the nonlinear frequency response characteristics, for both internal and external excitations. Transmission error due to variation in mesh stiffness was used as internal excitation and low frequency torque variations were used as external excitation. Two solution methods, digital simulation technique and the method of harmonic balance, were used to develop the steady state solutions for the internal sinusoidal excitation. Analytical predictions were shown to match satisfactorily with experimental data available in the literature. A parameter study showed that the mean load determined the conditions for no impacts, single sided impacts, and double-sided impacts. A six-degree-of-freedom model of a spur gear pair was developed by Torby [14]. The gears were supported by elastic bearings with viscous damping present. Varying mesh stiffness and friction in the gear mesh were included in the model. The equations of motion were solved by numerical integration.

Dynamic Models of Complete Gear Trains/Gear Boxes

Many researchers have modelled complete gearboxes in order to predict gear noise. Campell et al. [15] used finite element dynamic modelling methods to predict gear noise from a rear wheel drive automatic transmission. The model was used to investigate the effects of different component's inertia, stiffness, and resonance. The ring gear and shaft resonances and the tailstock housing stiffness were found to be significant design factors that influenced the gear-whine. Model construction issues were discussed as well as correlation of predicted gear noise traces with operating measurements. Ariga et al. [16] described a systematic approach to reduce the overall gear noise from a four-speed automatic transaxle in which the vibration characteristics were identified by finite element analysis [FEA]. A new gear train structure for effective in reducing gear noise was investigated. The effect of the modifications was verified experimentally, and the gear noise level was reduced substantially. Also changes in stiffness of the transmission case, at locations supporting the gear train bearings, were shown to affect the gear noise. Dynamic models of typical automotive gearing applications using spur, helical, bevel, hypoid, and planetary gear sets were developed by Donley et al. [17]. Basic formulations used in modelling different types of gears were discussed. These models were designed for use in finite element models of gearing systems for simulating gear-whine. A procedure for calculating the dynamic mesh force generated, and gear case response, per unit transmission error was proposed. A simplified automotive transmission was analysed to demonstrate the features of the proposed gear noise reduction technique. A basic approach to gearbox noise prediction was described by Mitchell and Daws [18]. The proposed method was dynamic modelling of the gearbox from inside out. The computational strategy for the determination of the dynamic response of the internal gearbox components was described. Force coupling between gears and dynamic coupling due to, for example, unbalance was discussed. The transfer matrix approach was used for the analysis and a benchmark example was presented to verify the calculation method. Hellinger et al. [19] used numerical methods to calculate gear noise from a transmission. They used finite element analysis to calculate natural frequencies and forced vibrations of the gearbox structure (housing). As input for the finite element calculation of forced response, they used the dynamic bearing forces of the shafts in the gearbox, calculated by multibody system software. Finite element analysis was used by Nurhadi [20] to investigate the influence of gear system parameters on noise generation. A direct time integrating method was used to predict sound generation, transmission, and radiation from mechanical structures. Naas et al. [21] optimized the gear noise from a car gearbox. They calculated transmission error and time varying stiffness of the gear mesh by using a finite element based computer program. The results were used as input to a torsional vibration model of the power train (from the engine to the wheels). The output from the torsional vibration model was mesh forces, which were transformed to the frequency domain and used as input to a finite element model of the gearbox, which was used to predict the forced response. As a result of the simulations, a modified gearbox was tested in a car and the gear noise was substantially reduced.

As per literature survey it is observed that a lot of instrument and measuring devices are required to control noise and vibration in gear boxes. Hence in order to investigate noise and vibration reduction in PGT in this paper a simple approach is proposed without the requirement of additional instruments like actuators, external power, and advanced signal processing techniques. This paper is organized as follows: Sections 1 and 2 focus on literature review and recent development in the subject, problem formulation, and objective of work is stated in Section 3; experimental set-up and measurement technique is explained in Section 4; Sections 5 and 6focus on basics of noise and vibration in PGT and results and discussion; concluding remark is explained in Section 7.

PROBLEM FORMULATION AND OBJECTIVE

Meshing of the gears in the planetary gear set that forms the ratio-changing mechanism of an automatic transmission produces gear noise over a wide range of driving conditions from low to high vehicle speed. As per the literature survey and industrial survey it is observed that the internal excitation caused by the variation in tooth mesh stiffness is a key factor in causing vibrations. To reduce gear vibrations, numbers of passive and active methods are reported. Many studies have been concentrated on the modification of gear teeth, but these methods have limitations on modifications. Passive methods like the use of periodic struts for gearbox support systems and periodic drive shafts are also reported to reduce gear vibrations. But these methods require additional actuators, external power, and many signal processing techniques.

Hence in order to investigate noise and vibration reduction in planetary gear train by phasing, in this work a simple approach is proposed without the requirement of additional instruments like actuators, external power, and signal processing techniques which generally results in increase in cost. To investigate noise and subsequent vibrations reduction in planetary gear set by phasing, two objectives were set such as design, development of schematic and systematic arrangement of system components to build test rig set-up and performance analysis by conducting trials with and without phasing.

EXPERIMENTAL SET-UP AND MEASUREMENTS

As per the requirement to accomplish the objectives it was necessary to develop the method to reduce PGT noise and vibrations by gear itself without requiring the additional energy, actuators, and advanced signal processing techniques. Viewing this need the method of noise reduction in planetary gears by phasing is introduced in this research work. In order to study the effect of phasing on noise and vibrations of planetary gear set the required experimental set-up was developed as shown in Figure 1.

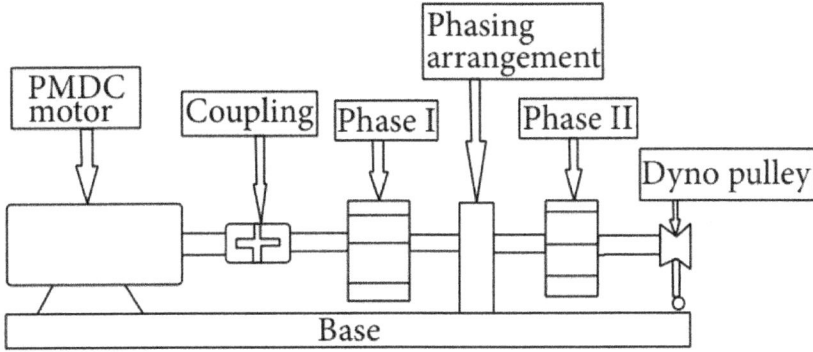

Figure 1: Schematic layout of test set-up for measurement of planetary gear set noise.

Figure 1 shows schematic layout of test set-up developed for the measurement of noise level of planetary gear set by phasing. Figure 2 also shows the position of various components like motor, planetary gear sets 1 and 2, coupling, and speed regulator. The experimental work was carried out to study the effect of meshing phasing on noise level and vibrations of Nylon-6 planetary gear set. For this purpose experimental set-up was built as shown in Figure 2. Rectangular plate is placed between planetary gear sets 1 and 2 to provide meshing phase difference between ring gear of gear sets 1 and 2. Noise level is measured for two different arrangements as with phasing and without phasing. Experimental set-up shown in Figure 2 consists of different components such as PMDC motor, love joy coupling, planetary gear set, and speed selector which are explained in this section.

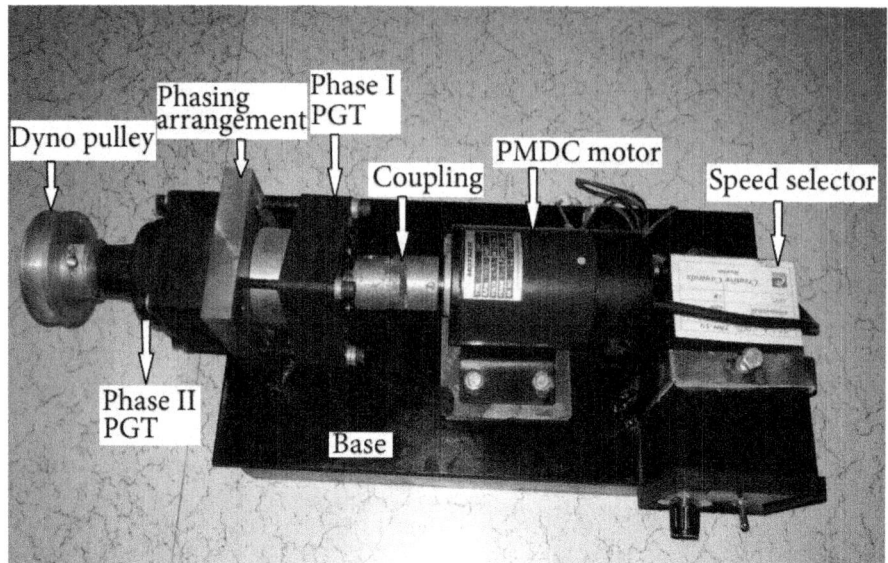

Figure 2: Experimental test rig.

Motor Selection

The speed reduction is to be achieved using two stage reductions; hence the stage wise reduction of the system is as follows.

Stage-I

(a) Input speed = 1440 rpm.

(b) Reduction ratio = 4.

(c) Output speed of stage I = 1440/4 = 360 rpm.

Stage-II

(d) Input speed = 360 rpm.

(e) Reduction ratio = 4.

(f) Output speed of stage-II = 360/4 = 90 rpm.

(g) Hence, maximum motor speed = 1500 rpm. Power = $2 \times \pi \times 1500 \times 0.5/60 = 78.5$ W.

Hence motor of 90 watt is selected as shown in Figure 3.

i. Motor type: fractional HP permanent magnet DC motor.

ii. Torque: 5 kgcm.

iii. Speed: 1500 rpm.

iv. Input power: 90 watt.

Figure 3: Permanent magnet DC motor.

Lovejoy Coupling

Lovejoy Coupling L-075 (as per specification shown in Table 1) is selected for the given application with outside diameter of hub (D_0) = 38 mm, inside diameter of hub (D_i) = 12 mm, Lovejoy Coupling L-075 considered to be a hollow shaft subjected to torsional load.

Table 1: Material selection for coupling

Designation	Ultimate tensile strength N/mm²	Yield strength N/mm²
EN 9	600	480

Selection of Gear Box

Nylon-6 planetary gear box (Table 2) is selected based on mechanical properties, wear resistance, lubrication and material availability, torque, and other parameters.

Table 2: Mechanical properties of Nylon-6 gear box

Mechanical properties	ASTM test method	Units	Nylon 6/6	Nylon 6/6 GF30
Tensile strength 73°F	D638	psi	12,400	27,000
Elongation 73°F	D638	%	90	3

Flexural strength, 73°F	D790	psi	17,000	39,100
Flexural modulus, 73°F	D790	psi	4.1×10^5	12×10^5
Izod impact strength, Notched, 73°F	D256	—	R120-M79	M101
Rockwell hardness	D785	ft-lbs/in.	1.2	2.1

Gear Pair-1 (Phase I). Tables 3 and 4 show the planet gear and internal gear ring specifications.

Table 3: Planet gear specification

Material	Nylon-6
Module	1.375 mm
Number of teeth	16
Addendum diameter	24.75 mm
Pitch circle diameter	22 mm

Table 4: Internal gear ring specification

Material	Nylon-6
Module	1.375 mm
Number of teeth	48
Addendum diameter	68.75 mm
Pitch circle diameter	66 mm

Gear Pair-2 (Phase I). Tables 3 and 5 show the planet gear and sun gear specification.

Table 5: Sun gear specification

Module	1.375 mm
Number of teeth	16
Addendum diameter	24.75 mm
Pitch circle diameter	22 mm

Coupler Shaft

Coupler shaft is used to connect planetary gear sets one and two. Table 6 shows different mechanical properties of selected material EN24.

Table 6: Selection of material for coupler shaft

Designation	Ultimate tensile strength N/mm²	Yield strength N/mm²
EN 24 (40 N; 2 Cr 1 Mo 28)	720	600

Coupler Shaft Bearing-I

Coupler shaft bearing is subjected to purely medium radial and axial loads; hence single row deep groove ball bearing is selected as per specifications shown in Table 7.

Table 7: Selection of coupler shaft bearing series 60

Bearing of basic design number (SKF)	D	D_1	D_2	D_3	B	Basic capacity	
6003	17	19	35	33	10	2850	4650

Gear Pair-3 (Phase II)

Tables 8 and 9 show the specification of planet gear and internal gear ring.

Table 8: Planet gear specification

Material	Nylon-6
Module	1.375 mm
Number of teeth	16
Addendum diameter	24.75 mm
Pitch circle diameter	22 mm

Table 9: Internal gear specification

Material	Nylon-6
Module	1.375
Number of teeth	48
Addendum diameter	68.75
Pitch circle diameter	66

Gear Pair-4 (Phase II)

Tables 10 and 11 show the planet gear and internal gear ring specification.

Table 10: Planet gear specification

Material	Nylon-6
Module	1.375 mm
Number of teeth	16
Addendum diameter	24.75
Pitch circle diameter	22

Table 11: Sun gear specification

Material	Nylon-6
Module	1.375 mm
Number of teeth	16
Addendum diameter	24.75
Pitch circle diameter	22

Sound Level Meter

For measurement of noise level in PGT sound level meter with specifications as shown in Table 12 was used.

Table 12: Specifications of sound level meter

Frequency range	31.5 Hz~8 KHz
Display	LCD
Measuring level range	35~130 dB
Accuracy	±1.5 dB (under reference conditions)
Dynamic range	65 dB
Power supply	One 9 V battery, 006P or IEC 6F22 or NEDA 1604
Calibration	Electrical calibration with the internal oscillator (1 kHz sine wave)

FFT Analyzer

For measurement of noise level in PGT FFT Analyzer with specifications as shown in Table 13 was used.

Table 13: Specifications of FFT Analyzer

Dynamic signal analyzer	Type PHOTON+
Electronics	Electronics differential amplifier, programmable gain amplifier,
Frequency range	Up to 84 kHz analysis frequency (192 k samples per second)
Voltage ranges	±0.01, ±0.1, ±1.0, ±10 V
Resolution	24-bit
Dynamic range	115 dBfs two-tone test, 100 linear averages
Accuracy	±0.04 dB (1 kHz sine at full scale)

Phasing Arrangement

Figure 4 shows planetary gear set used in experimental set-up. This planetary gear set is used to calculate the angle of indexing. Phase difference is provided in between ring gear of planetary gear sets 1 and 2. Figures 5(b) and 5(c) show without phasing and with phasing arrangement. Phase difference is provided as shown in Figure 5(a).

No of teeth of ring gear = 48.

Angle of pitch = 360/48 = 7.5° = 7.5/2 = 3.75°.

Angle of indexing = 5 × angle between two teeth = 5×3.75 = 18.75°.

Figure 4: Planetary gear set.

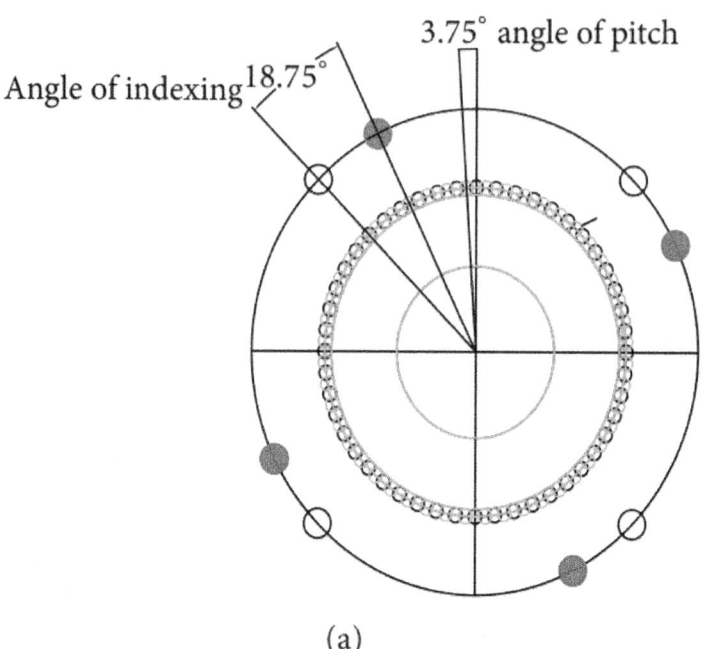

(a)

(b)

(c)

Figure 5: (a) Angle of Indexing, (b) gear arrangement without phasing, and (c) gear arrangement with phasing (normal gears).

NOISE MEASUREMENT IN PGT

Noise measurement and signal analysis are important tools when experimentally investigating gear noise. Gears create noise at specific frequencies, related to the rotational speed and number of teeth of the gear. It is also possible to detect different errors like for example, run out (eccentricity) due to side-band generation [22]. Closely related is also vibration measurement and signal analysis for the purpose of gear fault detection, used in machine diagnostics in order to detect gear failures before catastrophic failure occurs. Middelton [23] discussed noise testing of gearboxes in the production line in which a noise testing equipment, utilizing low cost digital analysis and control techniques, was described. For each gear of the gearbox, the speed was ramped up while measuring noise with three microphones. For each order of interest (gear mesh frequency and its harmonics) the pass/fail target levels were defined by testing a selection of gearboxes which had noise characteristics regarded as just acceptable. A test rig was developed by Gielisch and Heitmann [24] to investigate gear noise from a car rear axle, without the need for a complete vehicle. Vibrations were measured on the final drive casing and the corresponding forces and torques in the gearing were calculated. The investigation gave information about the dynamics of the driving gear and the possibility to make comparisons between different driving gears. Oswald et al. [25] investigated the influence of gear design on gearbox radiated noise in which nine different spur and helical gear designs were tested in a gear noise test rig to compare the noise radiated from the gearbox top for the various gear design and the results were summarized as follows.

- The total contact ratio was the most significant factor for reducing noise, increasing either the profile or face contact ratio reduced the noise.
- The noninvolute spur gears were 3-4 dB noisier than involute spur gears.
- High contact ratio spur gears showed a noise reduction of about 2 dB over standard spur gears.
- The noise level of double helical gears averaged about 4 dB higher than otherwise similar single helical gears.

In this paper method proposed in [7] is extended by applying novelty of phasing concept to reduce the noise level and resulting vibrations of planetary gear set without the requirement of additional instruments like actuators, external power, and signal processing techniques which generally results in increase in cost.

Figure 6 shows experimental set-up developed to measure noise of planetary gear set for with and without phasing arrangement. For development of test ring as shown in Figure 6 exhaustive literature review was carried out [22–26]. Specific motor speed (e.g., 1200 rpm) is selected with the help of speed regulator and gear set is rotated; then sound meter [26] and FFT Analyzer were used to measure noise level of gear set. In noise reduction tests, variation due to unintended effects, such as testing different part specimens or even reassembly with the same parts, may be of the same order of magnitude as the effect of deliberate design changes.

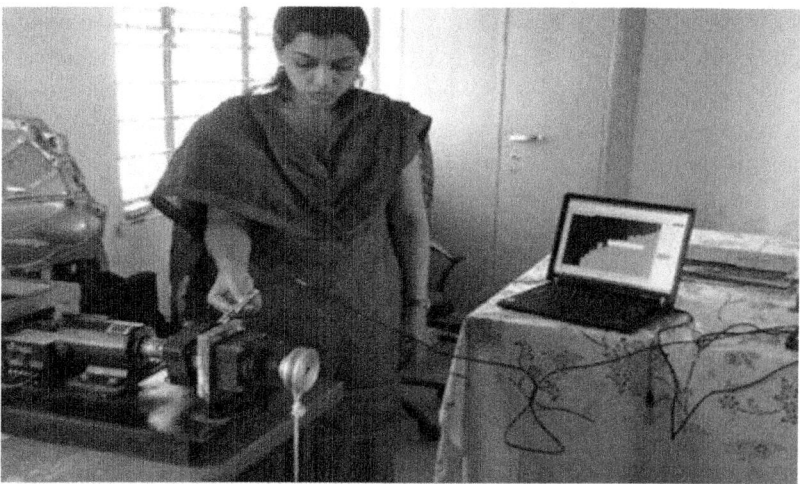

Figure 6: Recording results on sound meter and FFT spectrum analyzer.

RESULTS AND DISCUSSION

To validate the experimental findings number of trials were conducted by varying load from 1 Kg to 6 Kg and changes in torque, power, and resulting noise level are noted. Results obtained by without phasing and with phasing arrangement are shown in Tables 14 and 15. From Tables 14 and 15 relations between speed, measured torque, power, mechanical efficiency, and noise are plotted as shown in Figures 7, 8, and 9 for with and without phasing arrangement. The results were remarkably consistent with experimental measurements including excellent predictions of noise level by using sound level meter and FFT Analyzer.

Table 14: Observations without phasing arrangement

Load (Kg)	Speed (rpm)	Torque (Nm)	Power (W)	Efficiency (%)	Noise
1	74	1.56	12.00	17.14	78
2	73	3.13	23.35	33.31	80
3	72	4.70	35.52	50.74	83
4	71	6.27	46.71	66.73	86
5	71	7.84	58.29	83.93	89
6	70	9.41	69.03	98.90	91

Table 15: Observations with phasing arrangement

Load (Kg)	Speed (rpm)	Torque (Nm)	Power (W)	Efficiency (%)	Noise (dB)
1	75	1.56	12.25	17.50	72
2	74	3.13	24.25	34.64	74
3	73.5	4.70	36.17	51.67	76
4	72.5	6.27	47.60	68.00	79
5	72	7.84	59.11	84.44	82
6	71	9.41	69.96	99.45	85

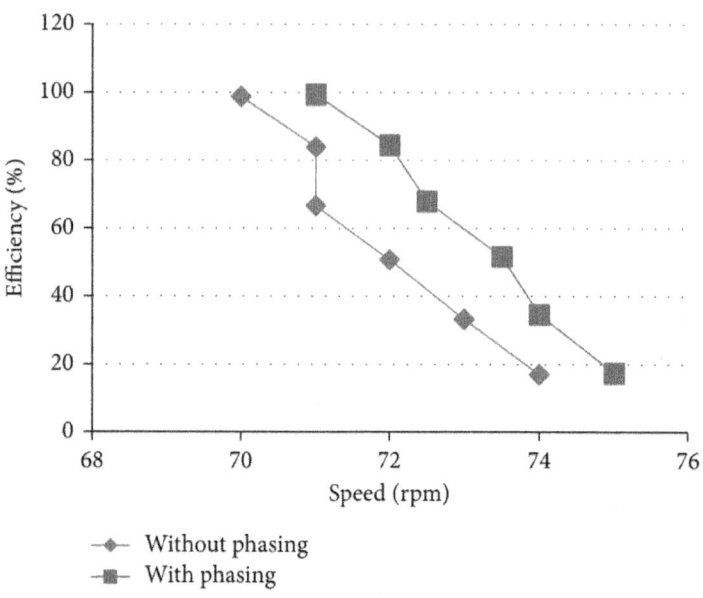

Figure 7: Comparison of speed (rpm) with efficiency (%) without and with phasing.

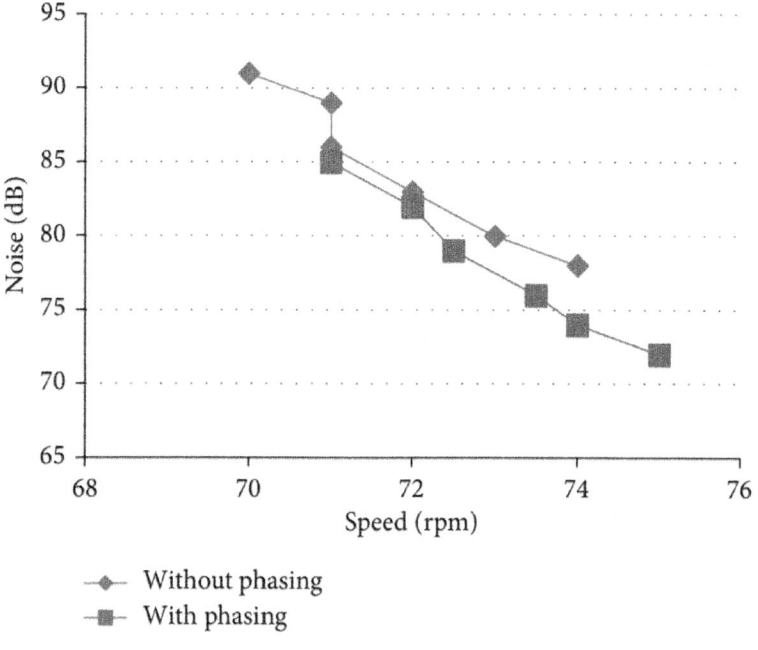

Figure 8: Comparison of speed (rpm) with noise (dB) without and with phasing.

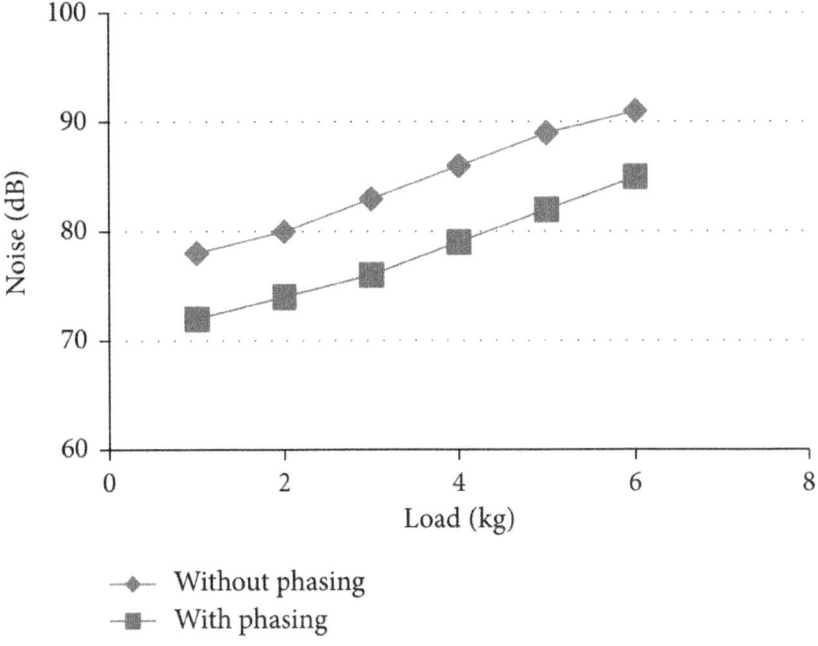

Figure 9: Comparison of load (kg) with noise (dB) without and with phasing.

From Figure 7 it is observed that as speed increases efficiency decreases. Efficiency of with phasing arrangement is greater as compared to without phasing arrangement at motor speed of 1200 rpm. Efficiency is maximum with 98.90% value at 70 rpm output speed at dyno pulley for without phasing arrangement, whereas efficiency is maximum as 99.45% value at 71 rpm output speed at dyno pulley with phasing arrangement. This shows efficiency is improved by using with phasing arrangement.

From Figure 8 it is observed that as speed increases, noise level decreases because speed decreases with increase in load. Noise level in without phasing arrangement (i.e., 78 dB at 74 rpm) is greater as compared to with phasing arrangement (i.e., 74 dB at 74 rpm) at motor speed of 1200 rpm. This shows noise level decreases with 4 dB by using with phasing arrangement. From Figure 8 it is concluded that the level of noise decreases when phase difference is provided. Figure 9 shows the effect of the meshing phase difference on the measured noise level. From Figure 9 it is seen that as load increases noise level increases drastically in case without phasing arrangement as compared to with phasing arrangement at motor speed of 1200 rpm.

Figures 10 and 11 show the variation of noise level with frequency at 1200 rpm for 6 Kg load without and with phasing obtained by FFT Analyzer.

By comparing Figures 10 and 11 it is seen that the average noise level is reduced by 6 dB to 7 dB. By using FFT Analyzer similar plots were obtained for load varying from 1 Kg to 5 Kg and resulting changes in noise level were recorded for required frequencies. For 1 kg load, from Figure 12 noise level with magnitude of 84 dB at frequency of 2000 Hz for without phasing arrangement is seen, while in with phasing arrangement noise measured found to be 77 dB. This indicates that the level of noise was reduced by 7 dB with phasing arrangement. For 2 kg load, from Figure 13 noise level with magnitude of 84.53 dB at frequency of 2000 Hz for without phasing arrangement is seen, while in with phasing arrangement noise measured found to be 76 dB. This indicates that the level of noise was reduced by 8.53 dB with phasing arrangement. For 3 kg load, from Figure 14 noise level with magnitude of 80 dB at frequency of 2000 Hz for without phasing arrangement is seen, while in with phasing arrangement noise measured found to be 76 dB. This indicates that the level of noise was reduced by 4 dB with phasing arrangement. For 4 kg load, from Figure 15 noise level with magnitude of 85 dB at frequency of 2000 Hz for without phasing arrangement is seen, while in with phasing arrangement noise level measured found to be 78 dB. This indicates that the level of noise was reduced by 7 dB with phasing arrangement. For 5 kg load, from Figure 16 noise level with magnitude of 84 dB at frequency of 2000 Hz for without phasing arrangement is seen, while in with phasing arrangement noise measured found to be 77 dB. This indicates that the level of noise was reduced by 7 dB with phasing arrangement. For 6 kg load, from Figure 17 noise level with magnitude of 85 dB at frequency of 2000 Hz for without phasing arrangement is seen, while in with phasing arrangement noise measured found to be 79 dB. This indicates that the level of noise was reduced by 6 dB with phasing arrangement.

Experimental Investigations of Noise Control in Planetary Gear Set by... 137

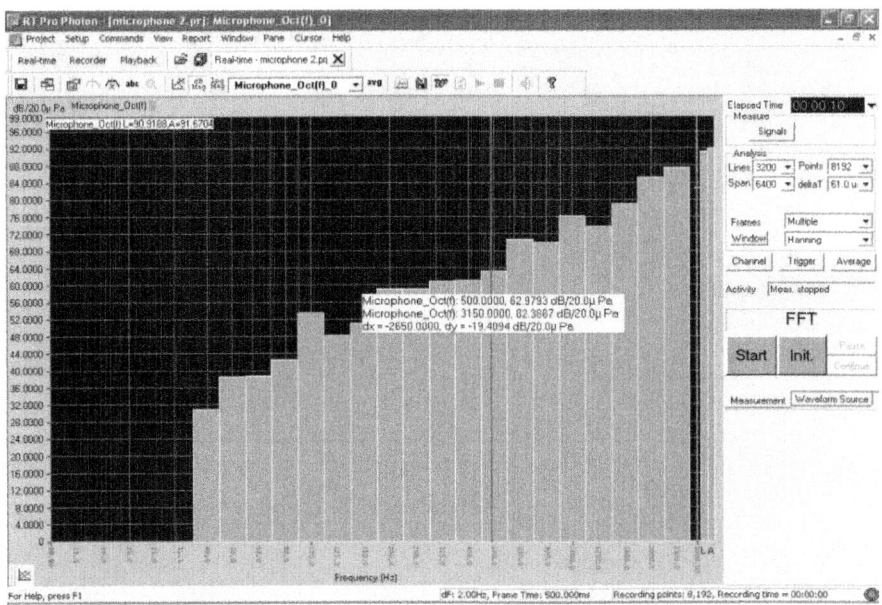

Figure 10: Variation of noise (dB) with frequency (Hz) at 1200 rpm for 6 Kg load without phasing.

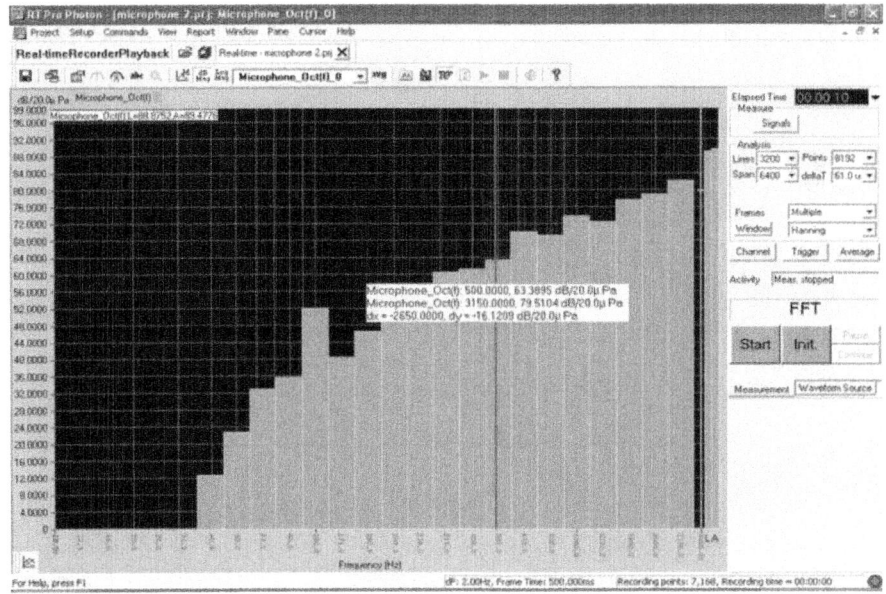

Figure 11: Variation of noise (dB) with frequency (Hz) at 1200 rpm for 6 Kg load with phasing.

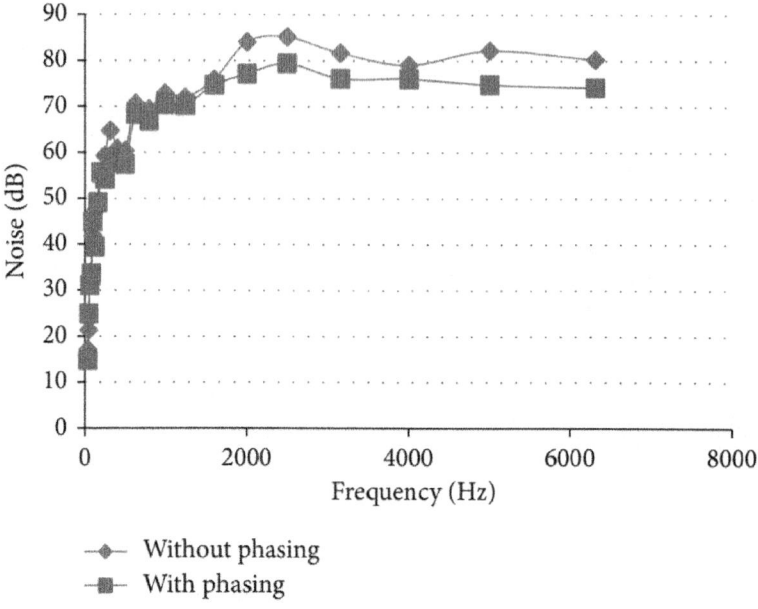

Figure 12: Variation of noise (dB) with frequency (Hz) at 1200 rpm 1 Kg load.

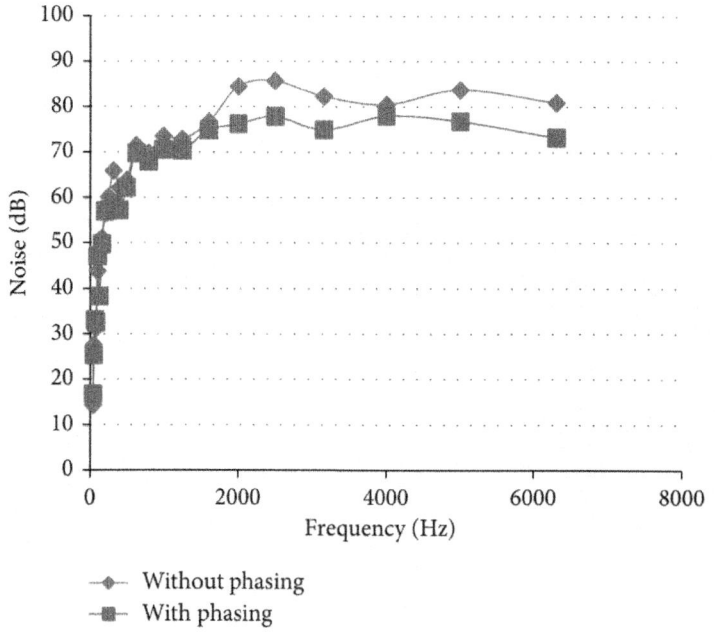

Figure 13: Variation of noise (dB) with frequency (Hz) at 1200 rpm at 2 kg load.

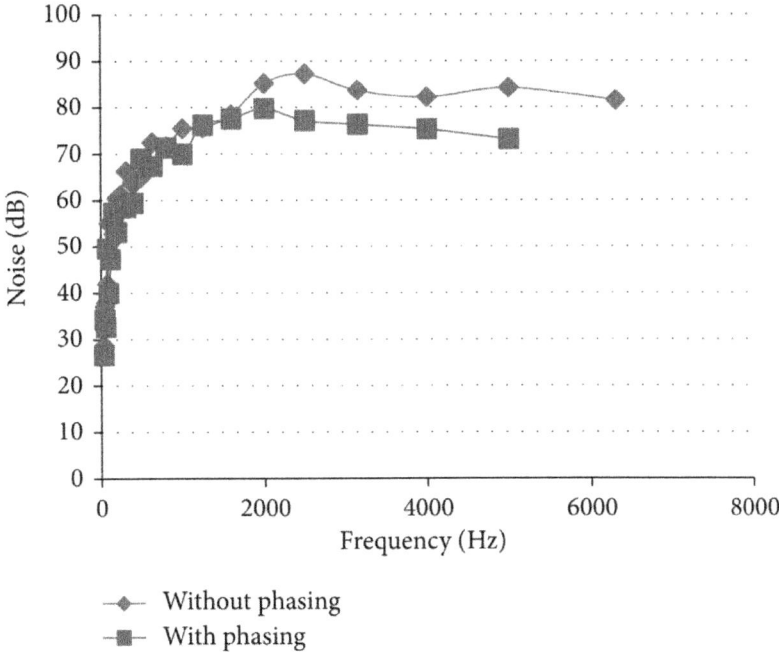

Figure 14: Variation of noise (dB) with frequency (Hz) at 1200 rpm at 3 kg load.

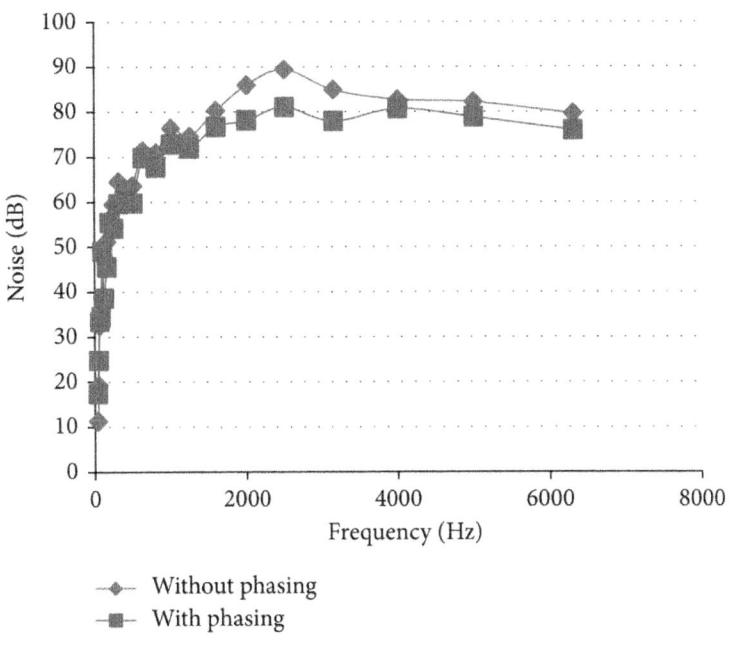

Figure 15: Variation of noise (dB) with frequency (Hz) at 1200 rpm at 4 kg load.

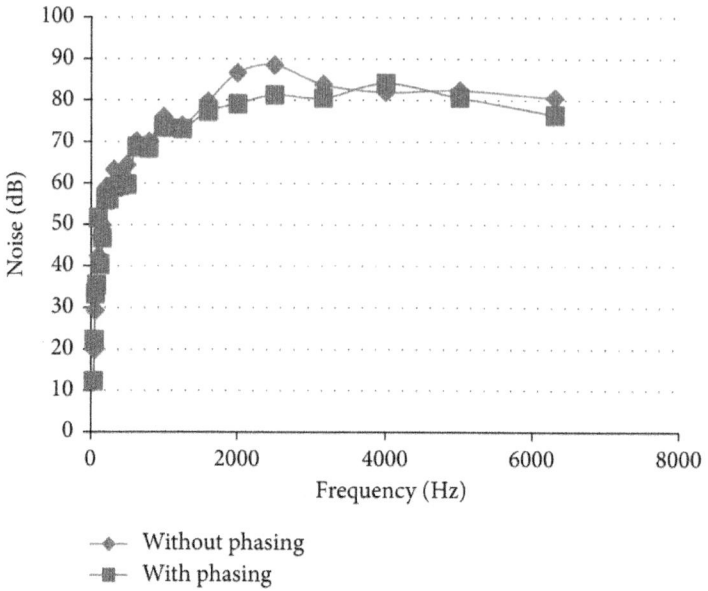

Figure 16: Variation of noise (dB) with frequency (Hz) at 1200 rpm at 5 kg load.

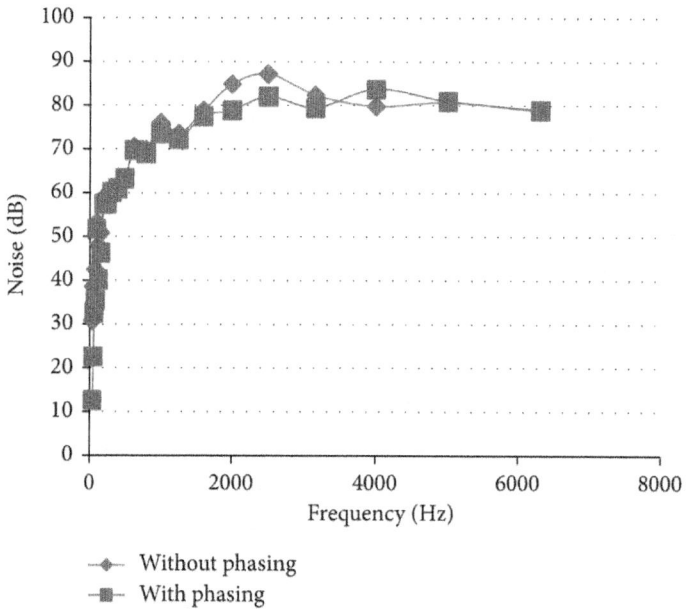

Figure 17: Variation of noise (dB) with frequency (Hz) at 1200 rpm at 6 kg load.

From these observations it is observed that the results obtained by sound level meter and FFT Analyzer are remarkably consistent with experimental measurements including excellent predictions of noise level as per literature survey.

CONCLUSION

The primary objective of this research work was to investigate noise reduction in planetary gear set by phasing. This objective was achieved with the help of extensive experimental investigations. Comparing results obtained by with and without phasing arrangement at 1200 rpm, it is observed that the level of noise in planetary gear set with meshing phase difference was approximately 6 dB to 7 dB lower than that of planetary gear set without a meshing phase. From the observations it is seen that the results obtained by sound level meter and FFT Analyzer are remarkably consistent with experimental measurements. So from the experimental results obtained by sound level meter and FFT Analyzer it is observed that by applying the meshing phase difference one can reduce planetary gear set noise and resulting vibrations.

REFERENCES

1. R. G. Schlegel and K. C. Mard, "Transmission noise control-approaches in helicopter design," inProceedings of the ASME Design Engineering Conference, ASME paper 67-DE-58, New York, NY, USA, 1967.
2. D. L. Seager, "Conditions for the neutralization of excitation by the teeth in Epicyclic Gearing," Journal of Mechanical Engineering Science, vol. 17, no. 5, pp. 293–298, 1975.
3. W. E. Palmer and R. R. Fuehrer, "Noise control in planetary transmissions," SAE Technical Paper770561, 1977.
4. A. Kaharamam and G. W. Blankership, "Planet mesh phasing in Epicyclic gear sets," in Proceedings of the International Gearing Conference, pp. 99–104, Newcastle, Wash, USA, 1994.
5. R. G. Parker, "Physical explanation for the effectiveness of planet phasing to suppress planetary gear vibration," Journal of Sound and Vibration, vol. 236, no. 4, pp. 561–573, 2000.
6. C. Gill-Jeong, "Numerical study on reducing the vibration of spur gear pairs with phasing," Journal of Sound and Vibration, vol. 329, no. 19, pp. 3915–3927, 2010.
7. Y. Chen and A. Ishibashi, "Investigation of noise and vibration of planetary gear drives," Gear Technology, vol. 23, no. 1, pp. 48–55, 2006.

8. R. G. Parker, V. Agashe, and S. M. Vijayakar, "Dynamic response of a planetary gear system using a finite element/contact mechanics model," Journal of Mechanical Design, Transactions of the ASME, vol. 122, no. 3, pp. 304–310, 2000.
9. P. Velex and L. Flamand, "Dynamic response of planetary trains to mesh parametric excitations,"Journal of Mechanical Design, vol. 118, no. 1, pp. 7–14, 1996.
10. R. J. Drago, "How to design quiet transmissions," Machine Design, vol. 52, no. 28, pp. 175–181, 1980.
11. H. N. Özgüven and D. R. Houser, "Mathematical models used in gear dynamics—a review," Journal of Sound and Vibration, vol. 121, no. 3, pp. 383–411, 1988.
12. W. Cheng, "Simulation of the stochastic vibration of spur gears," in Proceedings of the 16th International Conference on Computer Aided Production Engineering, Edinburgh, UK, November 1988.
13. A. Kahraman and R. Singh, "Non-linear dynamics of a spur gear pair," Journal of Sound and Vibration, vol. 142, no. 1, pp. 49–75, 1990.
14. B. Torby, Spur-Gear Dynamics, vol. 13 of TRITA-MMK, Royal Institute of Technology, Stockholm, Sweden, 1995.
15. B. Campell, W. Stokes, G. Steyer, M. Clapper, R. Krishnaswami, and N. Gagnon, "Gear noise reduction of an automatic transmission trough finite element dynamic simulation," SAE Technical Paper 971966, 1966.
16. K. Ariga, T. Abe, Y. Yokoyama, and Y. Enomoto, "Reduction of transaxle gear noise by gear train modification," SAE Technical Paper 922108, 1992.
17. M. G. Donley, T. C. Lim, and G. C. Steyer, "Dynamic analysis of automotive gearing systems," SAE Technical Paper 920762, 1992.
18. L. D. Mitchell and J. Daws, "A basic approach to gearbox noise prediction," SAE Technical Paper821065, 1982.
19. W. Hellinger, H. Raffel, and G. Rainer, "Numerical methods to calculate gear transmission noise," SAE Technical Paper 971965, 1965.
20. I. Nurhadi, Investigation of the influence of gear system parameters on noise generation [Ph.D. thesis], The University of Wisconsin, Madison, Wis, USA, 1985.
21. J. Naas, H. Stoffels, and A. Troska, "Integrierte berechnung zur optimierung des verzahnungsgeräusches eines Pkw-handschaltgetriebes," ATZ Automobiltechnische Zeitschrift, vol. 103, no. 1, pp. 16–17, 2001.
22. P. J. Sweeney, Transmission error measurement and analysis [Ph.D.

thesis], University of New South Wales, New South Wales, Australia, 1995.

23. A. H. Middelton, "Noise testing of gearboxes and transmissions using low cost digital analysis and control techniques," SAE Technical Paper 861284, 1986.

24. F. Gielisch and F. T. Heitmann, "Ermittlung der Geräuschanregungaus der Achsgetriebeverzahnungdurch Prüfstandsversuche," ATZ Automobiltechnische Zeitschrift, vol. 100, no. 4, pp. 282–286, 1998.

25. F. B. Oswald, D. P. Townsend, M. J. Valco, R. H. Spencer, R. J. Drago, and J. W. Lenski Jr., "Influence of gear design on gearbox radiated noise," Gear Technology, vol. 15, no. 1, pp. 10–15, 1998.

26. S. H. Gawande, S. N. Shaikh, R. N. Yerrawar, and K. A. Mahajan, "Noise level reduction in planetary gear set," Journal of Mechanical Design & Vibration, vol. 2, no. 3, pp. 60–62, 2014.

Chapter 8

NOISE SOURCE IDENTIFICATION OF A RING-PLATE CYCLOID REDUCER BASED ON COHERENCE ANALYSIS

Bing Yang[1] and Yan Liu[2]

[1]School of Mechanical Engineering, Dalian Jiaotong University, Dalian 116028, China

[2]School of Traffic and Transportation Engineering, Dalian Jiaotong University, Dalian 116028, China

ABSTRACT

A ring-plate-type cycloid speed reducer is one of the most important reducers owing to its low volume, compactness, smooth and high performance, and high reliability. The vibration and noise tests of the reducer prototype are completed using the HEAD acoustics multichannel noise test and analysis system. The characteristics of the vibration and noise are obtained based on coherence analysis and the noise sources are identified. The conclusions provide the bases for further noise research and control of the ring-plate-type cycloid reducer.

INTRODUCTION

Speed reducers are used in various fields for the purposes of speed and torque conversion. Speed reducers have many kinds such as worm reducer, crane reducer, cycloid reducer, planetary gear reducer, and ring-plate-type reducer. A ring-plate-type cycloid speed reducer is one of the most important reducers owing to its low volume, smooth and high performance, and high reliability. The internal transmission structure of the ring-plate-type cycloid speed reducer is shown in Figure 1 [1–3]. The input shaft equipped with a driving involute gear is supported by the reducer. Two driven gears are mounted on two driven cranks. Four ring plates with pin gears are connected to the two driven cranks. Two cranks have the same length. Thus, ring plate and two cranks become a parallel four-bar mechanism. When the input shaft rotates, it causes the two cranks to rotate. The four ring plates mounted to the cranks rotate. Then the cycloidal gear rotates. The output shaft rotates since the cycloidal gear is mounted to the output shaft [4].

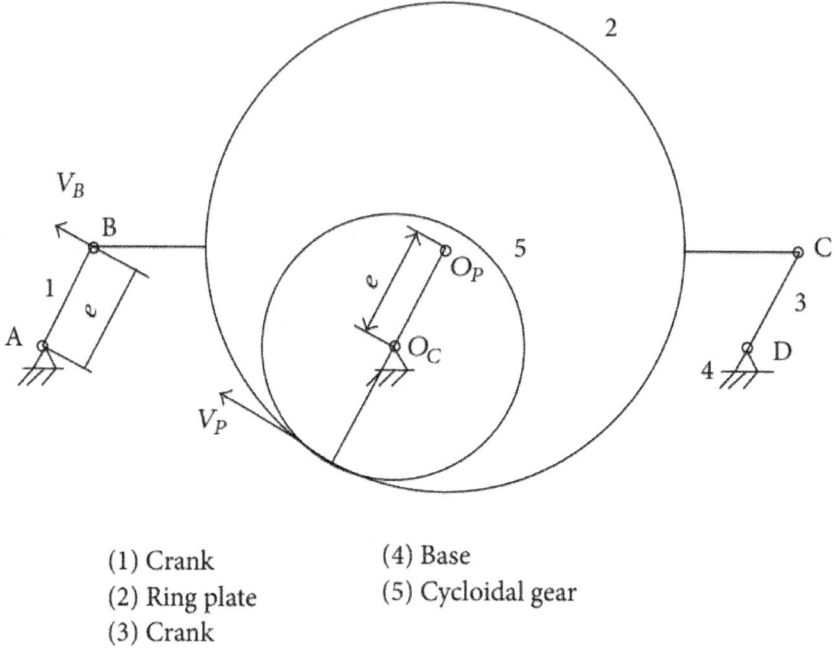

(1) Crank
(2) Ring plate
(3) Crank
(4) Base
(5) Cycloidal gear

Figure 1: Crank ring-plate-type cycloid speed reducer.

At present, the application of the reducer is limited to some extent because of the noise level. So, it has practical significance to research the vibration and noise of the reducer [5–10]. The vibration and noise tests of the reducer are completed using the HEAD acoustics multichannel noise test and analysis system. The characteristics of the vibration and noise are obtained based on coherence analysis and the noise sources of the double crank ring-plate-type pin-cycloid gear reducer are identified.

NOISE SOURCE IDENTIFICATION METHODS

Mechanical equipment noise control is mainly in three areas: sound source control, transmission route control, and recipient protection. And the control of the noise source is the most fundamental and effective method. The premise is identifying the main sources of the equipment [11–13]. There are many noise source identification methods such as subjective evaluation, respectively run, and lead cover. The following are frequency spectrum and coherence analysis methods.

Frequency Spectrum

The data measured are time-domain signal in general. In order to obtain the frequency characteristics of the noise source, frequency spectrum analysis is often made. The frequency spectrum can be generated via a Fourier transform of the signal, into a single harmonic component to study, to thereby obtain the frequency structure of the signal, and the amplitude and phase information of the harmonic and the resulting are usually presented as frequency spectrum diagrams [14–16]. The peak values of the frequency spectrum diagrams are closely but not necessarily related to the main source of noise. A peak in the noise and vibration frequency spectrum diagram may come from several noise sources, and sometimes a noise source may produce more than one peak in the frequency spectrum diagram.

Coherence Analysis

A coherence function is the description of the relevance of the two signals in a system [17, 18]. Let x(t) and y(t) be the signals; the Autocorrelation function is defined as

$$R_x(\tau) = E\left[x(t+\tau)\,x(t)\right] \tag{1}$$

The cross-correlation function is defined as

$$R_{xy}(\tau) = E\left[x(t+\tau)\,y(t)\right]. \tag{2}$$

The correlation coefficient between (t) and $y(t)$ can be expressed as

$$\rho_{xy} = \frac{\sum_{t=0}^{\infty} x(t)\,y(t)}{\left[\sum_{t=0}^{\infty} x^2(t)\,\sum_{t=0}^{\infty} y^2(t)\right]^{0.5}}. \tag{3}$$

The inputs of a system are the noise or vibration; the only linear output is that (t). (t) is the sum of the linear outputs. The coherence function between an input and the output can be expressed as

$$\gamma_{iy}^2(\omega) = \frac{\left|G_{iy}(\omega)\right|^2}{G_{ii}(\omega)\,G_{yy}(\omega)}. \tag{4}$$

VIBRATION AND NOISE TESTS

Test Equipment

The vibration and noise tests of the reducer prototype are completed using the HEAD acoustics test and analysis system in an ordinary laboratory. The

HEAD acoustics multichannel noise test and analysis system is selected for the measurement. The system is made of SQLab II data acquisition recorder, G. R. A. S. microphones and KISTLER acceleration sensors, and Artemis software. The data collection and analysis process of the HEAD acoustics multichannel noise test and analysis system is shown in Figure 2. The signals are collected from reducer through sensors to the front end. The analog signal is converted into a digital signal through SQLab. At last, the digital signals are analyzed by Artemis software with different analysis methods.

Figure 2: Signal collection and analysis process.

Test Point Position

The ring-plate cycloid reducer noise comes mainly from structural noise. Structural noise is generated by the imbalance, the mechanical collision, and structural resonance and propagates to the space propagation through the shaft, the bearing, and the block. Structural noise mainly comes from the following two sections. The first, section, machinery parts produces sound when they rotate. Such as shaft, gear, motor and so on. The other section comes from components engagement and sound.

The main noise sources of the ring-plate cycloid reducer are the following aspects through experiences:

- the noise generated by ring plates and cycloidal gear when they mesh,
- the noise generated by involute spur gears when they mesh,
- the noise caused by driving motor, and
- the noise caused unbalanced installation of ring plates.

According to the possible noise sources, three vibration test points and one noise test point were positioned in the test procedure Figure 3. The location of the test points is shown in Figure 4. The detailed information of test points is shown in Table 1.

Table 1: Test point location explanation

Point name	Test point location
1	Vibration test point, near the input shaft, on the surface of the reducer

2	Vibration test point, near the cycloid gear, on the surface of the reducer
3	Vibration test point, near the output shaft, on the surface of the reducer
5	Noise test point, 1 meter above the reducer

Figure 3: Test points position.

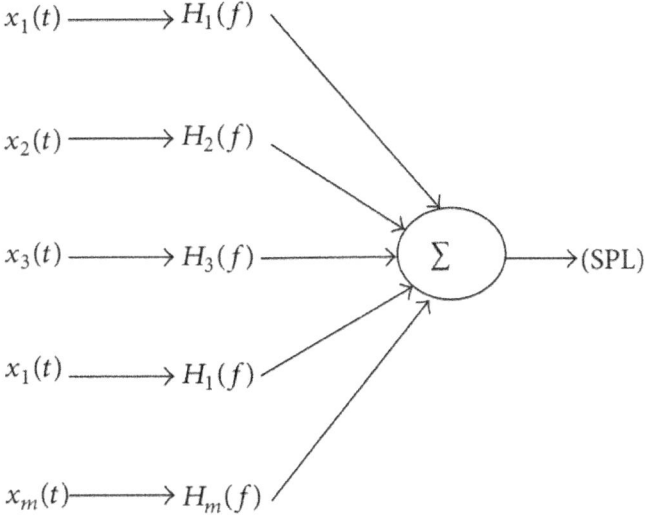

Figure 4: Multi-input single-output linear system.

NOISE SOURCE IDENTIFICATION

The double-crank four ring-plate-type cycloid reducer internal-noise-source-related parameters can be expressed as

$$k = \frac{f}{f_s}, \tag{5}$$

where f is the main frequency and f_s is the shaft frequency, and the shaft frequency can be expressed as

$$f_s = \frac{n}{60}, \tag{6}$$

where n is the rotation speed of the gears.

The data collected from three vibrations test points are supposed as $x_1(t)$, $x_2(t)$, $x_3(t)$, and the signal received from the noise test point is $y(t)$. The multi-input single-output linear system model is shown in Figure 4. The collected vibration acceleration signals $x1(t)$, $x_2(t)$, $x_3(t)$ are the inputs of the system; the collected sound pressure level signal $y(t)$ is the output of the linear outputs. The four test point data are collected synchronously through the front end. The collected data are analyzed through the Artemis software.

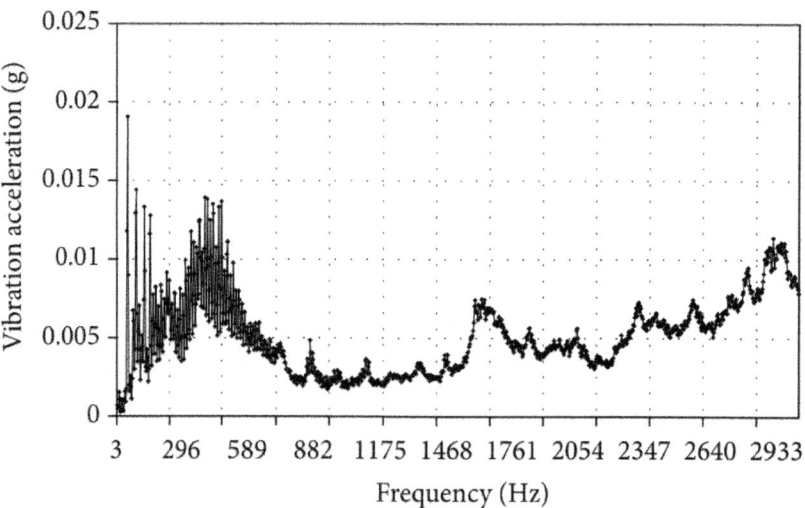

Figure 5: Vibration acceleration of test point 1.

In order to grasp the noise distributing laws of the reducer, three different speeds and three different loads were selected. They are 750 rotations per minute, 1000 rotations per minute, and 1250 rotations per minute. The three different loads are 40%, 60%, and 80% of the full load. Figure 5 shows the vibration acceleration spectrum diagram of test point 1 with the input shaft rotation of 1250 rpm. The frequency spectrum diagrams of the other speeds are not given here due to limited space.

Figure 5 shows that the vibration acceleration of test point 1 shakes a little bit in the low frequency range. Figure 6 shows the noise frequency spectrum diagram of test point 5 with the input shaft rotation of 1250 rpm. Figure 6 shows that the sound pressure lever of test point 5 is 86.4 dB. The frequencies corresponding to the four peaks are 584 Hz, 1168 Hz, 1744 Hz, and 2360 Hz, respectively.

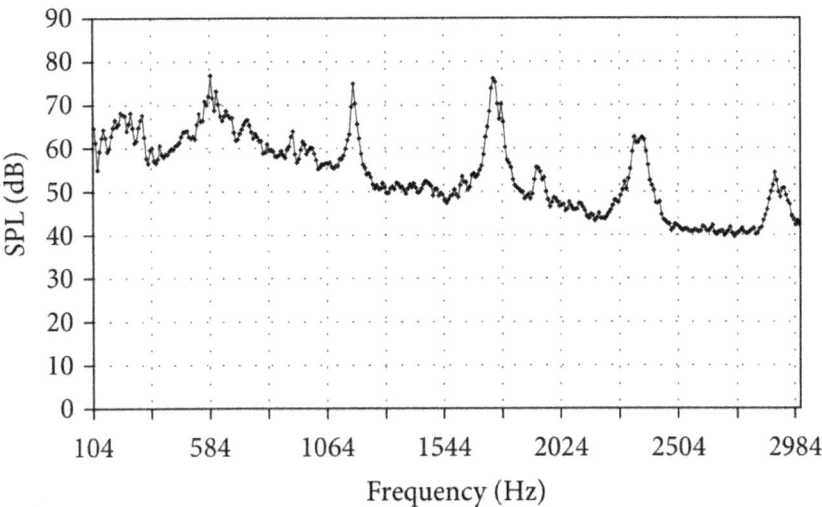

Figure 6: SPL of test point 5.

Figure 7 shows the Autocorrelation frequency spectrum diagram of test point 1. The frequencies corresponding to the three peaks are 584 Hz, 1768 Hz, and 2320 Hz, respectively. Figure 8 shows the Autocorrelation frequency spectrum diagram of test point 2. The frequencies corresponding to the peaks are 1752 Hz and 2336 Hz.

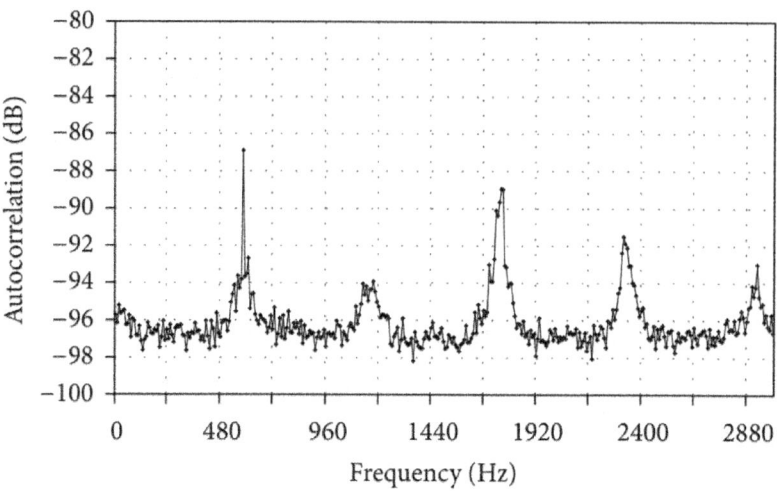

Figure 7: Autocorrelation of test point 1.

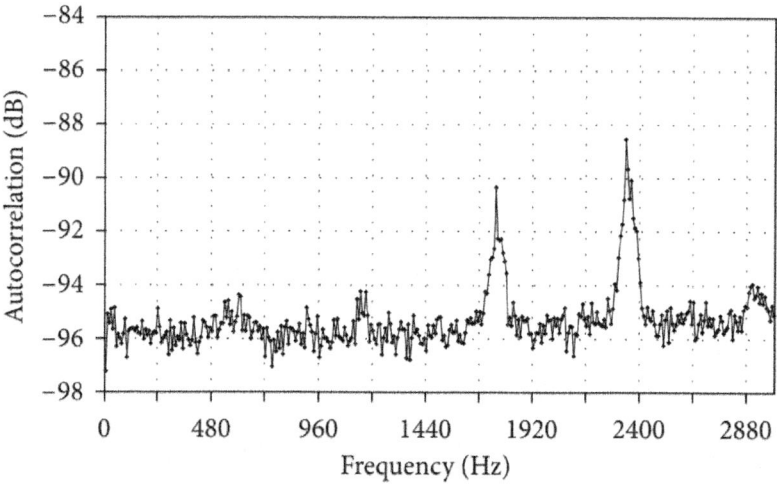

Figure 8: Autocorrelation of test point 2.

Figure 9 shows the cross-spectrum analysis diagram of between test point 5 and test point 1. Figure 10 shows the cross-spectrum analysis diagram of between test point 5 and test point 2. From the analysis we know the following conclusions. The contribution of the noise from the position of the ring plates is the biggest. The involute spur gears' contribution to the noise is not too high. So the noise reduction measures should be taken.

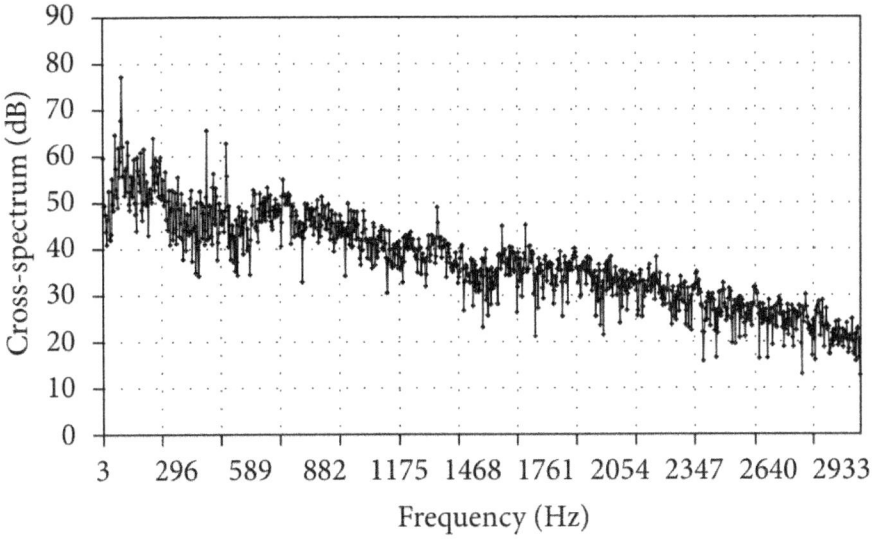

Figure 9: Cross-spectrum between 5 and 1.

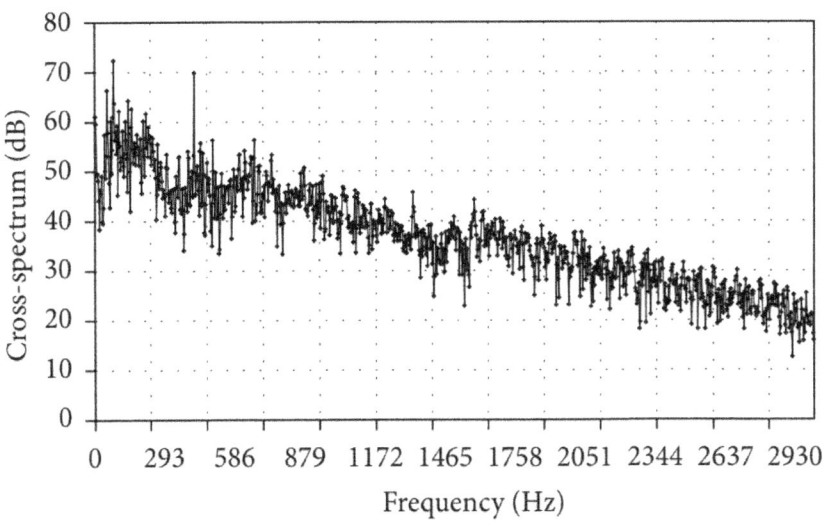

Figure 10: Cross-spectrum between points 5 and 2.

CONCLUSIONS

Double-crank four ring-plate cycloid speed reducer is one of the most important reducers owing to its low volume, smooth and high performance, and high reliability. In this paper, the characteristics of the vibration and noise are obtained after the vibration and noise tests. The noise sources are identified as the ring plates. To reduce the noise of the reducer, the structure of the ring plates or the unbalance of installation may be taken into consideration.

ACKNOWLEDGMENT

This work is supported by Key Laboratory of Modern Acoustics, Ministry of Education, China (Project no. 1108).

REFERENCES

1. W. D. He, X. Li, and L. Li, "Study on double crank ringplate-type cycloid drive," Journal of Mechanical Engineering, vol. 36, no. 5, pp. 84–88, 2000.
2. W. D. He, X. Li, L. Li, and B. Wen, "Optimum design and experiment for reducing vibration and noise of double crank ring-plate-type pin-cycloid planetary drive," Jixie Gongcheng Xuebao/Journal of Mechanical Engineering, vol. 46, no. 23, pp. 53–60, 2010.
3. W. D. He, Q. Lu, et al., "Study on pin cycloid planetary gear reducer used in propeller pitch variator for 1.5 MW wind turbine," Journal of Mechanical Transmission, vol. 36, no. 7, pp. 1–4, 2012.
4. X. Li, W. D. He, L. Li, and L. C. Schmidt, "A new cycloid drive with high-load capacity and high efficiency," ASME Journal of Mechanical Design, vol. 126, no. 4, pp. 683–686, 2004.
5. N. B. Roozen, J. Van den Oetelaar, A. Geerlings, and T. Vliegenthart, "Source identification and noise reduction of a reciprocating compressor; A case history," International Journal of Acoustics and Vibrations, vol. 14, no. 2, pp. 90–98, 2009.
6. F. Payri, A. Broatch, X. Margot, and L. Monelletta, "Sound quality assessment of Diesel combustion noise using in-cylinder pressure components," Measurement Science and Technology, vol. 20, no. 1, Article ID 015107, 2009.
7. P. Simon, "Retrieving the three-dimensional matter power spectrum and galaxy biasing parameters from lensing tomography," Astronomy and Astrophysics, vol. 543, no. 1, p. 1, 2012.
8. C. C. Zhu, D. T. Qin, and S. Hong, "Study on surface noise distribution of three-ring reducer," Journal of Chongqing University, vol. 4, pp. 18–21, 2000.
9. T. J. Lin, Y. J. Liao, R. F. Li, and W. Liu, "Numerical simulation and experimental study on radiation noise of double-ring gear reducer,"

Journal of Vibration and Shock, vol. 29, no. 3, pp. 43–47, 2010.

10. J. Prezelj and M. Čudina, "Quantification of aerodynamically induced noise and vibration-induced noise in a suction unit," Proceedings of the Institution of Mechanical Engineers C, vol. 225, no. 3, pp. 617–624, 2011.

11. Z. Chu, W. Wang, and X. Xiao, "Noise source identification and noise abatement of truck under idle speed condition," Transactions of the Chinese Society of Agricultural Engineering, vol. 26, no. 5, pp. 153–158, 2010.

12. A. Ramachandran, M. C. Reddy, and R. Moodithaya, "Minimization and identification of conducted emission bearing current in variable speed induction motor drives using PWM inverter," Indian Academy of Sciences, vol. 33, no. 5, pp. 615–628, 2008.

13. R. S. Magalhães, C. H. O. Fontes, L. A. L. Almeida, and M. Embiruçu, "Identification of hybrid ARXneural network models for three-dimensional simulation of a vibroacoustic system," Journal of Sound and Vibration, vol. 330, no. 21, pp. 5138–5150, 2011.

14. X. D. Wu, S. G. Zuo, and Y. Lu, "Identifying noise source based on application to partial coherence analysis in fuel cell car," Journal of Noise and Vibration, vol. 3, pp. 81–84, 2008.

15. M. L. Chen and S. M. Li, "Coherence functions method for signal source identification," China Mechanical Engineering, vol. 1, pp. 95–100, 2007.

16. K. S. Oh, S. K. Lee, and S. J. Kim, "Identification and reduction of noise from axles in a passenger vehicle," Noise Control Engineering Journal, vol. 56, no. 5, pp. 332–341, 2008.

17. F. Deželak, "Effective noise control through identification and ranking of incoherent noise sources,"Noise Control Engineering Journal, vol. 58, no. 2, pp. 212–221, 2010.

18. H. E. Camargo, P. A. Ravetta, R. A. Burdisso, and A. K. Smith, "Application of phased array technology for identification of low frequency noise sources," Journal of Low Frequency Noise Vibration and Active Control, vol. 28, no. 4, pp. 237–244, 2009.

Chapter 9

NOISE OF INDUCTION MACHINES

Marcel Janda, Ondrej Vitek and Vitezslav Hajek

Brno University of Technology, Czech Republic

INTRODUCTION

Diagnostics of electric machines is very interesting and extensive. There are many methods used to detect properties of electrical machines. Between diagnostic methods include too the measurement and analyze of noise, which generates electrical machines.

Itself the noise of electric machines is by product of the machine operation. The generation of noise is involved in many physical principles.

Noise of electrical machinery is generated by the vibration of machine parts. Gradual spread of vibration from the engine to the surroundings causes pulses of air with certain frequencies. This creates a sound wave generator, which can be within a certain frequency range, audible to humans.

The main sources of noise in electrical machines are time change of the electromagnetic fields, noise of bearings and other mechanical sources. Finally, the unwanted noise is creating too due to coolant flow or parts that come into contact with coolant in electric machines. Level of noise sources in electrical machines depends on the structural arrangement and the accuracy of engine design. A major problem in measurement noise is interference environment. For a perfect suppression ambient noise is necessary to have a specialized laboratory. It should be also measured machine have isolated from vibration, which it may be transferred from storage.

BASIC CONCEPTS OF NOISE

The sound wave is generated by vibrating objects and can be defined as mechanical interference with the finite speed of advancing through the media.

These waves have small amplitude, adiabatic oscillation are characterized by a wave speed, wavelength, frequency and amplitude. Sound has the character of longitudinal sound waves in the direction of propagation in the environment. In other words, it is the movement of individual particles of the medium in a direction parallel to the transmission of energy. Sound waves spread in three-dimensional environment from the source. It is same in all directions, if is the environment homogeneous. Sound waves can be polarized, they cannot have orientation. Non-polarized waves can oscillate in any direction in the plane perpendicular to the direction of propagation.

Sound amplitude can be measured as sound pressure level (SPL), sound intensity (SIL), sound power level (SWL) and the intensity of the acoustic energy (SED).The human ear can perceive sound waves of sufficient intensity and frequencies are ranging from 20 to 20,000 Hz. The Minimum sound intensity is different for different frequency and it is called the threshold of audibility. Range of sound intensity, which can capture the human ear, is 10-12 to 1 W/m2W/m2corresponding sound pressure of 20 MPa. Maximum sound level in which humans feel pain is called threshold of pain. Amplitude of sound about pressure 100 Pa is very loud.

Acoustic Pressure

Concentration of particle of vibrating environment corresponds with increase or decrease pressure inside gasses and liquids. This means that the total pressure in the environment is changing and therefore fluctuates around the initial static value or barometric pressure. The acoustic pressure is then considered deviation of the total pressure from the static pressure. For acoustic pressure is valid relationship

$$p_c = p_b + p(t) \tag{1}$$

$$p(t) = p_0 \cdot \cos(\omega \cdot t + \varphi) = p_0 \cdot \cos(2 \cdot \pi \cdot f \cdot t + \varphi) \tag{2}$$

$$p_c = p_0 + p_0 \cdot \cos(2 \cdot \pi \cdot f \cdot t + \varphi) \tag{3}$$

Where
- p_c... Acoustic Pressure [Pa]
- p_b... Barometric pressure [Pa]
- p_0...Amplitude of sound pressure [Pa]
- f... Frequency [Hz]
- t... Time [s]
- φ...phase shift

For effective sound pressure value is valid relationship

$$p_{ef} \frac{p_0}{\sqrt{2}} \qquad (4)$$

Acoustic pressure is a variable and it describing the noise source quantitatively. The measured level depends on the observer's distance from the source and the quality of the transmission environment Acoustic pressure level gives us information on the total sound pressure across a entire audible band. For sound pressure level is valid relationship

$$L\_p = 20.\log(p/p\_0) \qquad (5)$$

Where

- p.... Static pressure [Pa]
- p_0...Minimum value of static pressure, which is able to capture the human ear [Pa]

Sound Power Level

Mechanical vibrations are transmitted in form of mechanical energy from the source through acoustic waves. Sound power level is called the energy that passes per unit time over surface. For sound output, we can write the relationship

$$P_{ac} = p_c . v . A \qquad (6)$$

Where

P_{ac}... Sound power level [W]

p_c.... Acoustic pressure [Pa]

v....Vibration velocity of particles[m/s]m/s

A...Area

The sound power level depend on the the environment parameters and distance from the measurement point. The sound power level can be expressed as

$$L_{Pac} = 20.\log\left(\frac{v}{10^{-9}}\right) + k \qquad (7)$$

Where

k... constant

Acoustic Intensity

Acoustic intensity is a vector quantity that describes the amount and direction of flow of acoustic energy in the environment. Vector of acoustic intensity is time-change of instantaneous sound pressure and it is corresponding instantaneous speed of vibrating particle environment in the same place

$$I = \overline{p(t).v(t)} \tag{8}$$

Where

I... Acoustic intensity [W/m²]

NOISE SOURCES

From the physical point of view, mechanical sound is waves in a flexible environment. The Frequency range of sound audibly for human ear is from 20 Hz to 20 kHz. The sound spreads in all directions from resources by transmitting acoustic wave energy. Division by frequencies of sound waves:
- Infrasound - up to 20Hz
- Low frequency - 20Hz to 40Hz
- RF - 8kHz to 16kHz
- Ultrasound - 20kHz over

Dividing the sound by timing:
- Steady
- variable
- intermittent
- pulse

The interest noise frequency is over 1000 Hz for induction machines. Noise of Electrical Machines is characterized as a set of sounds that are caused by rapid changes in air pressure. These changes cause most commonly:
- Vibration of machine parts or the whole of its surface
- Aerodynamic phenomena that lead to pulsation of pressure near the machine

Basic sources of noise are induction motors (see diagram):
- Electromagnetic source
- The mechanical source
- The aerodynamic source

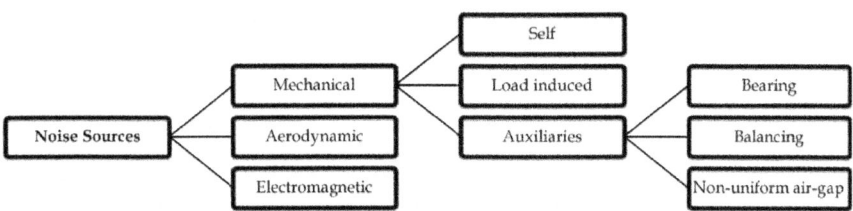

Figure 1. Division of noise sources in electrical machines

The noise from electromagnetic source is the most typical component noise of electrical machine. Its cause is the vibration of motor body, or other parts of the machine on which work the electromagnetic forces. Frequency Spectrum noise of the electromagnetic source has discrete character, while there is very distinct directional radiation characteristics of this component in many cases.

Determining the influence of this component on the overall noise of electric machine is often simply done so, that after switching off the machine from the network is observed decline in the acoustic signal in time. If is this decline immediately, then it is obviously a component of the noise of electromagnetic origin. Another method of investigation is the measurement of electromagnetic noise spectrum for different values of power - or even frequency.

Noise origin of ventilation is crucial observe especially in machines with high rotational speed. Detailed analysis of the fan noise shows that the main source in this case is very fan with its nearest surroundings. It is the decisive exceeds other sources of noise, which can be, for example rotor wings, radial or axial cooling channels in the machine, input and output caps and the like.

Frequency analysis of noise ventilation origin shows that the spectrum has a broadband character, either discrete or vice versa. In the first case, the aerodynamic noise is created from turbulent airflow near fan blade and near the entrance, but also the output edges of blades. These pulsations are uneven both in space and in time, so the frequency spectrum created of wind noise is broadband and contains all components of the audible band.

In contrast, discrete nature of the spectrum, sometimes the siren phenomenon can arise. This phenomenon arise if the fan or behind obstacles (such as a blade with these obstacles) is not profile of velocity uniform air flow around the wheel circumference, leading to periodic pulsation of pressure. Then the siren noise is produced naturally.

The noise of mechanical origin is primarily inflicted on roller bearings and unbalance of rotating machine parts. Rolling bearings can create multiple frequency components, which have their origin mainly in inequality as part of rolling paths of the bearing rings. In principle, the noise of mechanical origin has a mixed character.

Electromagnetic Noise

The influence of magnetic induction in the air gap formed magnetic forces; these forces operate across various directions. They may also have various amplitude and frequency. Their work is split between the rotor and stator of electric machine. Their characteristics depend on the size and shape of the air gap and a number of other factors.

The construction of the rotor is the main radiator noise machine. If the frequency is close to the radial force or equal to one natural frequency of the stator system, resonance occurs which leads distorted stator system with vibration and noise. Magnetostriction noise electric machine can be neglected in most cases due low and high frequency 2f arrangement r = 2p of radial forces, where f is the fundamental frequency and p is the number of pole pairs. However, the radial forces due magnetostriction can reach up to 50% the radial forces produced in the air gap magnetic field.

Magnetic flux density wave

Stator: $B_{m1}.\cos(\omega_1.t+k.\alpha+\Phi_1)$ (9)

Rotor: $B_{m2}.\cos(\omega_2.t+l.\alpha+\Phi_2)$ (10)

Where

- B_{m1}...Amplitude of magnetic flux density in stator [T]
- B_{m2}...Amplitude of magnetic flux density in rotor [T]
- $\omega\Phi_1$...Angular frequency of stator magnetic fields
- $\omega\Phi_1$...Angular frequency of rotor magnetic fields
- k,l...Variable (values 1,2,3,4,....)

For total wave of magnetic flux density can be write relationship

$P_{mr} = 0,5. B_{m1}. B_{m2}. \cos[(\omega_1 + \omega_2).t + (k+l).\alpha + (\Phi_1 + \Phi_2)]+$

$+0,5. B_{m1}. B_{m2}. \cos[(\omega_1 + \omega_2).t + (k-l).\alpha + (\Phi_1 - \Phi_2)]$ (11)

The magnetic stress wave has worked in radial directions on the stator and on active surfaces of rotor. This causing the deformation and subsequently cause the vibration and noise.

The mixed product of stator and rotor winding space harmonic create forces at frequencies

$$f_r = f_1 \cdot \left[\frac{n.Z_r}{p} \cdot (1-s) + 2\right]$$

$$f_r = f_1 \cdot \left[\frac{n.Z_r}{p} \cdot (1-s)\right] \tag{12}$$

Where
- f_1... Supply frequency [Hz]
- n.... value n=0, ±1, ±2,... [-]
- p... number of pole pairs [-]
- N_{rs}...Number of rotor slots [-]
- s... slip

The mixed product of stator winding and rotor eccentricity space harmonics create forces with frequencies

$$f_r = f_1 \cdot \left[\frac{n.N_{rs}}{p} \cdot (1-s) + 2\right]$$

$$f_r = f_1 \cdot \left[\frac{n.N_{rs}}{p} \cdot (1-s)\right]$$

$$f_r = f_1 \cdot \left[\frac{n.N_{rs}}{p} \cdot (1-s) + \frac{1-s}{p}\right]$$

$$f_r = f_1 \cdot \left[\frac{n.N_{rs}}{p} \cdot (1-s) + 2 + \frac{1-s}{p}\right] \tag{13}$$

The mixed product of stator winding and rotor saturation harmonics create forces at frequencies

$$f_r = f_1 \cdot \left[\frac{n.N_{rs}}{p} \cdot (1-s) + 4\right] \tag{14}$$

$$f_r = f_1 \cdot \left[\frac{n.N_{rs}}{p} \cdot (1-s) + 2\right] \tag{15}$$

Rotor Eccentricity

The air gap width depends only on position (no on time) in the static eccentricity. We conclude that the magnetic field in the air gap is rotating synchronous speed. That is given by the mains frequency and with the number of pole pair's induction machine. Modulation of magnetic field in one period is function, which is represented by a variable air gap, i.e. a function of its conductivity.

Static eccentricity is defined as the rotor axis offset from the axis of the stator. The air gap has a variable character. There is stronger interaction of stator and rotor magnetic field at the point where the gap is smaller. Influence of the static eccentricity manifests as the emergence of side frequency bands, which are shifted from the mains frequency f_1 of the synchronous frequency f. For static eccentricity is the angular frequency $\Omega_\varepsilon = 0$.

Static eccentricity is straight-line. The frequency for static eccentricity is twice power frequency

$$f_{stat} = 2.f_1 \qquad (16)$$

The relative eccentricity ε is defined as

$$\varepsilon = \frac{e}{g} = \frac{e}{R-r} \qquad (17)$$

Where
- R... Inner stator core radius
- r... Outer rotor radius
- e... Rotor eccentricity
- g... Ideal uniform air-gap for e=0

Dynamic eccentricity occurs when the rotor failure or its affiliates. Ratios are complicated by the fact that the width of air gap is not just a function of position, but is also a function of time. The variable air gap is changing at the rotation of the rotor. There is emergence of side bands that appears in the frequency range of vibrations of electric machine.

Angular frequency for dynamic eccentricity

$$\Omega_\varepsilon = \Omega.(1-s) = \frac{\omega}{p}.(1-s) = 2.\pi.\frac{f}{p}.(1-s) \qquad (18)$$

The frequency generated by the dynamic eccentricity

$$f_{DYN} = f_1 \pm (1-s).f_{SO} \qquad (19)$$

For frequency generated by eccentricity is true also relationship

$$f_{exc} \left[(n_{rt}.R \pm n_d).\frac{1-s}{p}.n_{\omega s} \right].f \qquad (20)$$

Where
- R...Number of grooves engine
- s... Chute
- p... Number of pole pairs

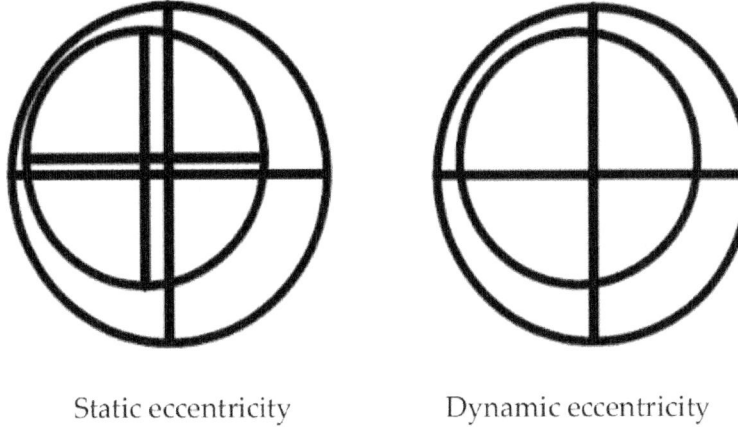

Static eccentricity   Dynamic eccentricity

Figure 2. Rotor eccentricity

Aerodynamic Noise

Aerodynamic noise arises most often around the fan, or in the vicinity of the machine that behaves like a fan. Noise can be created too on the necks stator slot windings or rotor. The aerodynamic noise sources can also include the noise produced by air flow inside and outside the design of electrical machines.

The main reason for the fan noise is formation of turbulent air flow around the blades. This noise is characterized by spectrum in a wide range, which has continuous character. Acoustic performance is increasing with the square of velocity. Siren noise can be eliminated by increasing the distance between the impeller and the stationary obstacle.

For the fan noise can write the relationship

$$L_A = 60.\log U_2 + 10.\log D_2.b_2 + \sum k_1 \qquad (21)$$

Where
- U_2...Outer speed of fan on the circuit [m.s^{-1}]
- D_2...Outer diameter of the fan [m]
- B_2... Fan width [m]
- k_1... Constants for the correction

The vortex frequency is expressed by

$$f_v = 0,185.\frac{v}{D_2} \qquad (22)$$

The frequency of the pure tone due to the fan blades is given by relationship

$$f_f = N_b \cdot \frac{N}{60} \qquad (23)$$

Where

- N...speed [rev/min.]
- N_b... Number of fan blades [-]

Sound power level of aerodynamic noise is

$$L_W = 67 + 10 \cdot \log_{10}(P_{out}) + 10 \cdot \log_{10}(p) \qquad (24)$$

$$L_W = 40 + 10 \cdot \log_{10}(Q) + 20 \cdot \log_{10}(p) \qquad (25)$$

$$L_W = 94 + 20 \cdot \log_{10}(P_{out}) - 10 \cdot \log_{10}(Q) \qquad (26)$$

Where

- P_{out}... Motor rated power [kW]
- p... Fan static pressure [Pa]
- Q...Flow rate [m³.s⁻¹]

Reducing aerodynamic noise in electrical machines can be use the following ways:

- Reducing the required amount of coolant used for ventilation of electrical machines
- Optimal design of fan. Especially the number and shape of the fan blades has an impact on the noise generated by the electric machine.
- To minimize the noise is needed to prevent vibration machine parts, which come into contact with a cooling medium.

Mechanical Noise Sources

Mechanical noise is mainly due with bearings, their defects, ovality, sliding contacts, bent shaft, rotor unbalance, shaft misalignment, couplings, U-joints, gears etc. In principle, the mechanical source of noise has a mixed character. The noise caused by unbalance of rotating parts and noise of bearing is spread after machine constructions very well. Dynamic balancing in production serves to reducing the noise of mechanical source. Especially for machines with high speed is necessary to perfect balance. Also, compliance with the manufacturing tolerances and technological processes, especially in the manufacture of small machines is the best solution to reduce the noise of mechanical source. Any change in noise from this source can mean failure of the mechanical parts inside the motor. For example, the bearings failure (damaged ball) is appear in the noise spectrum. There are specific frequencies by individual damage. The very faults of bearings and their effect on the noise spectrum of the electric machines are now well mapped.

Design of bearings can be either a sliding or rolling bearings. Rolling bearings can create multiple vibration frequencies, which have their origin mainly in the uneven parts or rolling themselves paths to the bearing rings. If bearing has mechanical damage, there is uneven movement of the whole system and thus increasing vibration and noise of the electric machine.

The main mechanical sources of the noise
- Alignment
- Inaccurate machining of parts
- Running speed
- Number of rolling elements carrying the load
- Mechanical resonance frequency of the outer ring
- Lubrication conditions
- Temperature

Rolling Bearings

The noise of rolling bearings depends on the type of bearing and its construction and accuracy of bearing parts. The increase in vibration and noise level of bearings, when the rotational speed changes from n_1 to n_2 can be expressed as

$$\Delta L_v = 20 \cdot \log \frac{n_2}{n_1} \qquad (27)$$

Ball pass frequency – outer race

$$f_{or} = \frac{N_b}{2} \cdot n_m \cdot \left(1 - \frac{d_b}{D} \cdot \cos\alpha\right) \qquad (28)$$

Where
- D...Pitch diameter [m]
- n_m... Rotation speed [rev/s]
- N_b...Number of balls [-]
- d_b...Diameter of balls]m]

- α...Contact angle of balls

Ball pass frequency – inner race

$$f_{ir} = \frac{N_b}{2} \cdot n_m \cdot \left(1 + \frac{d_b}{D} \cdot \cos\alpha\right) \tag{29}$$

Sleeve Bearings

- Uneven journal

$$f_{ov} = k \cdot n_m \tag{30}$$

k=1, 2, 3, ...

Axial groves

$$f_{gr} = N_g \cdot n_m \tag{31}$$

N_g...number of groove

Load Induced Noise

In certain cases, the vibrations and thus noise transmitted from the load, which is connected to the induction motor. In most cases, this occurs with wrong balance or bad connects of couplings. Uneven distribution load acting on the motor shaft or inappropriate use of gears may also affect noise machine. The only possible protection against these effects is the perfect balance of the whole set and if possible an even distribution of forces acting on the connecting elements. Noise arises too due to coupling of the machine with a load, e.g., shaft misalignment. Next noise arises from belt transmission, from cogwheels and couplings. It may also arise to noise due to mounting the machine on foundation or other structure.

NOISE MEASUREMENT

For measurement noise of induction machines can be used several techniques. The basic method for the measurement noise is the sound meter. It is a device which measures sound pressure.

Measurement Process

Measurement of noise can be divided into three main parts. The first part is data capture. For this purpose, the most commonly used microphones, or specialized equipment to measure noise (sound level meter). Their output is usually an analog signal, which must be further processed. When choosing of microphone is needed careful heed on certain parameters that can affect measurement accuracy. One of the most important parameter is the sensitivity of frequency. Worse microphones not recorded of the entire spectrum of the measured noise. Thanks to this complicates achieve it of accurate analysis results. Other parameters include the microphone sensitivity, which indicates the size of the output voltage (mV / Pa), depending on the pressure acting on the membrane. In addition, the structural dimensions of the measurement microphone and also the type of sound field that which is measured. Computers are most frequently use for Signal processing. For this reason it is necessary to convert from analog signal to digital form.

Large numbers of types A/D converters is on the market. Some are stand-alone converters; others are integrated to the specialized measurement cards. In both cases, the measurement depends on the three main parameters. The first is the measuring range of the converter. It gives the minimum, respectively maximum, measurable value. Because the signal is weak from a microphone, there should be used an amplifier for its amplification. Another parameter of the A/D converter is the bit depth conversion. This parameter defines the limitations of this device.

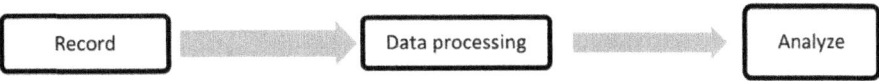

Figure 3. Block diagram of measurement process

Factors to selecting a suitable type of microphone are as follows

Table 1. Selecting factors of microphone

Characteristics of sound field	**Required accuracy**	**Environmental conditions**
Freely field for a closed chamber	Tolerance sensitivity	Noise level background
An important range of sound pressure levels	Frequency distortion tolerance	Humidity
An important frequency range	Phase distortion tolerance	Atmospheric pressure
	Tolerance of non-linear distortion	wind
	Own noise tolerance	Strong electromagnetic fields
		Mechanical shock

Sound Level Meter

Sound level meter is an essential instrument for measuring sound pressure levels. This device consists of the following components: Microphone, preamp, overload detector, central Unit, weighing Network, filters, amplifier, RMS detector, Output and Display.

One of the basic parameters of sound level meter is range of frequency. The sound intensity I has broad frequency range. The dispersion of the frequencies is from lower f_1 to higher f_2. The immediate value is indicated by I (f). For sound intensity is valid the relationship

$$I = \int_{f_1}^{f_2} I(f)df \qquad (32)$$

Where $I(f) = \frac{\Delta I}{\Delta f}$ is intensity in the frequency interval f = 1 Hz.

Spectral intensity level (ISL) L_{Is} is defined

$$L_{IS} = 10 log \left[\frac{I(f)}{I_{ref}}\right] \qquad (33)$$

Where I_{ref} is the reference intensity levels (for air $\frac{10^{-12}W}{m^2}$).

$$L_I = L_{IS} + 10 log(\Delta f) \qquad (34)$$

Similarly, the sound pressurse level L_p is related to the level of spectral noise L_{ps} as follows:

$$L_p = L_{ps} + 10 log(\Delta f) \qquad (35)$$
$$\Delta f = f_u - f_l \qquad (36)$$

Where f_l and f_u are the lower and upper frequency to half power.

FAST FOURIER TRANSFORMATION (FFT)

Fast Fourier Transformation is one of the most common mathematical functions, which is used for noise analysis of electrical machines. The Fast Fourier Transformation is applied in an increasing scale in science, engineering, and technology. The use of complex exponentials has often been convenient rather than fundamental. Most signals and functions used in real applications are real rather than complex. In areas such as digital filtering, convolution, correlation, image processing, and partial differential equations, the actual signals or functions, are real, but they are considered to be the real part of a complex quantity in order to be able to use the complex formulation of Fourier series and transforms. The complex Fourier transform (CFT) of a signal x(t)$-\infty \leq$ t

$\leq -\infty$ with finite energy, is defined as

$$x_c(f) = \int_{-\infty}^{\infty} x(t).e^{-j2.\pi.f.t} dt \tag{37}$$

The inverse complex Fourier transform (ICFT) is given by

$$x(t) = \int_{-\infty}^{\infty} x_c(f).e^{j.2.\pi.f.t} dt \tag{38}$$

The real Fourier transform (RFT) of x(t)xtcan be defined as

$$x(f) = 2\int_{-\infty}^{\infty} x(t).\cos(2.\pi.f.t + \Theta(f)) dt \tag{39}$$

Where:

$$\Theta(f) = \begin{cases} 0, & f \geq 0 \\ \frac{\pi}{2}, & f < 0 \end{cases}$$

The inverse real Fourier transform (IRFT) is given by

$$x(t) = \int_{-\infty}^{\infty} x(f).\cos(2.\pi.f.t + \Theta(t)) df \tag{40}$$

Equation (3) and (5) can be written for f≥0 as follows

$$x_1(f) = 2\int_{-\infty}^{\infty} x(t).\cos(2.\pi.f.t) dt \tag{41}$$

$$x_0(f) = 2\int_{-\infty}^{\infty} x(t).\sin(2.\pi.f.t) dt \tag{42}$$

and

$$x(t) = \int_{-\infty}^{\infty} [x_1(t).\cos(2.\pi.f.t) + x_0(t).\sin(2.\pi.f.t)] df \tag{43}$$

Thus x(f) equals x_1(f) for f≥0, and x_0(f) for f<0. x_1(f) and x_0(f) will be referred to as the cosine and the sine parts. The relationship between the CFT and the RFT can be expressed for f≥0 as $x_c(0) = x1(0)$

$$\begin{bmatrix} x_c(f) \\ x_c(-f) \end{bmatrix} = \frac{1}{2}.\begin{bmatrix} 1 & -j \\ 1 & j \end{bmatrix}.\begin{bmatrix} x_1(f) \\ x_{10}(f) \end{bmatrix}, f \neq 0 \tag{44}$$

Equation (44) reflects the fact that x_1(f) and x_0(f) are even and odd functions, respectively.

The inverse of (44) for x_1(f) and x_0(f) is

$$\begin{bmatrix} x_1(f) \\ x_0(-f) \end{bmatrix} = \begin{bmatrix} 1 & 1 \\ 1 & -j \end{bmatrix}.\begin{bmatrix} x_c(f) \\ x_c(-f) \end{bmatrix} \tag{45}$$

Equations (44) and (45) are very useful to convert from one representation to the other. When x(t) is real, x_1(f) and x_0(f) are also real. Then, (44) shows that x_c(f) and x_c(-f) are complex conjugates of each other. Equations (44) and (45) are also valid in the case of the discrete time Fourier transformation. In addition, they are valid for Fourier series and the discrete Fourier transforms with the replacement of f by the frequency index n. The RFT relations given by (43) can be proven by using (44), and writing (38) as

$$x(t) = \int_0^\infty \frac{1}{2}.[x_1(f) - jx_0(f)].e^{j.2.\pi.f.t}df + \int_0^{-\infty} \frac{1}{2}.[x_1(-f) - jx_0(-f)].e^{j.2.\pi.f.t}df \qquad (46)$$

then

$$x(t) = \int_0^\infty [x_1(f).\cos(2.\pi.f.t) + x_0(f).\sin(2.\pi.f.t)]df \qquad (47)$$

MEASUREMENT NOISE OF INDUCTION MACHINES

Disturbed Surroundings

Surrounding noise sources have an impact on the measurement of electrical machinery. It is not always possible to perform measurements in specialized laboratories, which are perfectly sound-insulated. To laboratory measurement can penetrate the noise from nearby sources (see Fig. 4), which is inaudible to the human ear. The interference from other sources can be created undesirable frequencies in the frequency band.

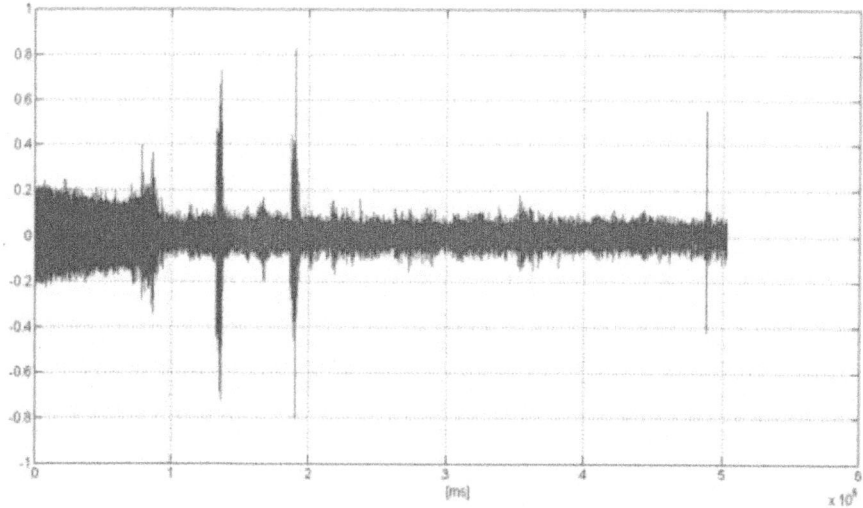

a) Noise detected in the laboratory

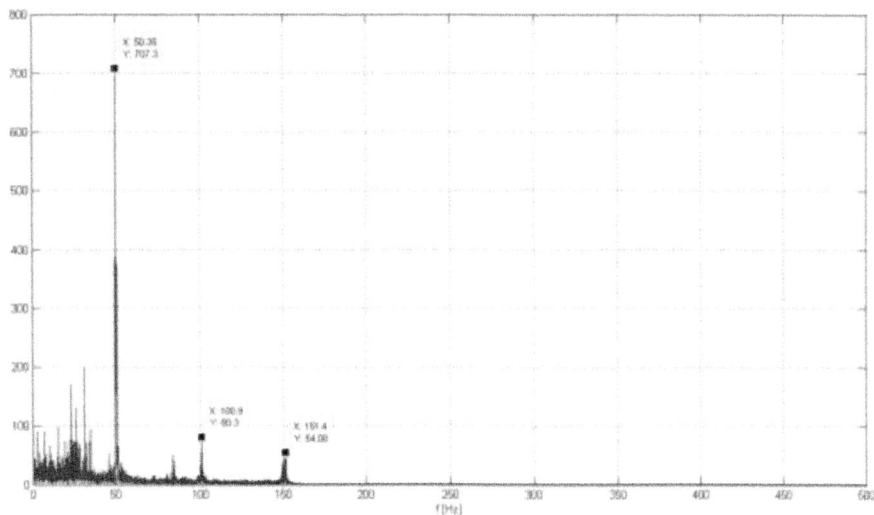

b) Fast Fourier Transformation of laboratory noise

Figure 4. Noise measurement in the laboratory when the machine is switched off.

Interference of other sources in the neighborhood of workplace cannot be directly prevented, but you can minimize their impact on analysis of the measured signal. Before the measurements it must be made measurement ambient noise before the main measurements. It is necessary to determine whether the background noise is random, or it is periodically repeated. In the case of random noise is preferable to wait to other time of measurement or it must count with errors in the measurement. In the event that can be measurement of noise repeated. Can be recorded the extent of spectral interference with which will be calculate when evaluating the measured results. From Spectral analyses of interference is possible to determine the proportion of individual harmonics. These harmonic then they can be the "subtracted" from the noise levels of electrical machines.

The next part of the measurement was performed on the induction motor which worked without a load. The electric motor was loosely placed on a foam board. This board was for suppression the transmission of vibrations from the surroundings. External vibrations are not desirable for accurate measurements. Measurement noise of electric machine, that is run, is shown in Fig. 5. As seen from the measured values, that the noise level is constantly fluctuating.

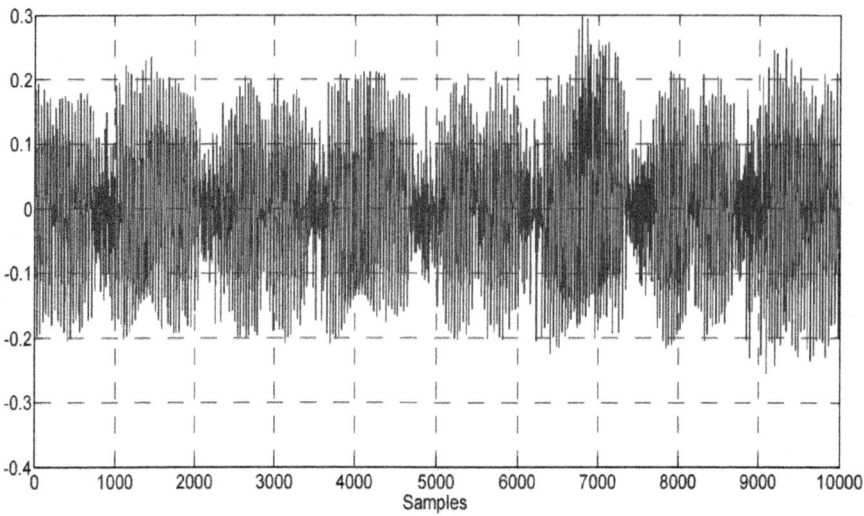

Figure 5. Noise of induction machines - no load

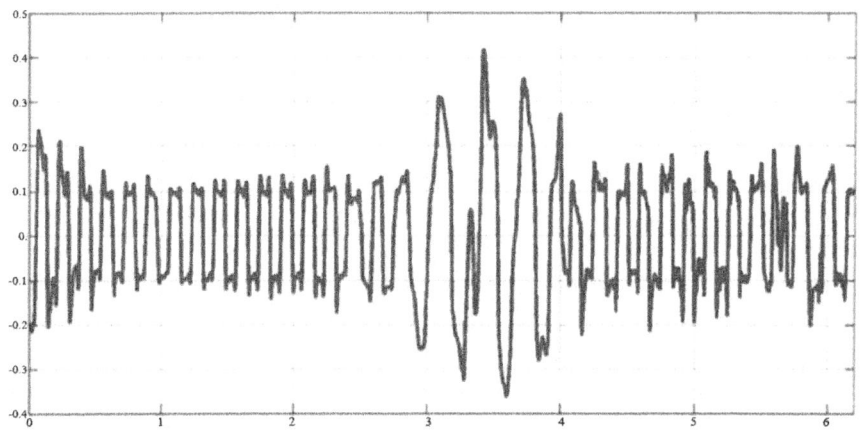

Figure 6. Noise of induction machine – 1 rotation.

Noise of Induction Machine

On Fig. 7is an analysis of the measured noise using MATLAB. Specifically, was carried Fast Fourier Transform (FFT). Dominant frequency is 600 Hz. This frequency is multiple of power supply frequency. It is a frequency of radial forces. In measurement signal can be involved many harmonics frequencies of radial forces. Than we can write equation

$$f_v = 6.k.f \tag{48}$$

Where

- f_v Frequency of radial force [Hz]
- f   Power supply frequency [Hz]
- k   Number (k=1, 2, 3, ...)

For f=50Hz are frequencies of radial forces f_v=300, 600, 900,...Hz.

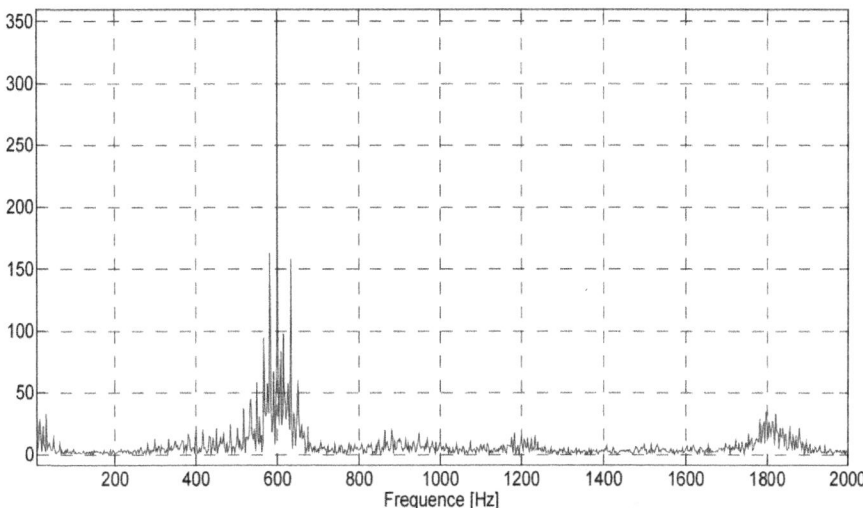

Figure 7. Fast Fourier Transformation of induction machine noise.

It was done measurements eccentricity of rotor. Eccentricity of rotor is shown in Fig. 9. From the measured values it was found that the largest deviations occur in the range of approximately 120 degrees.

When comparing the noise of induction machines recorded on one rotation and values of rotor eccentricity can see a connection. In both cases (Fig. 8 and Fig. 9) appeared larger deflection in the range of 120 degrees. Extreme deviation is in a different quadrant in each graph. This is due to the different measurement principles. Noise measurements done digitally, while measuring the eccentricity was used mechanical method. It was therefore not possible to accurately determine the initial rotor position in both measurements.

it can be argued that the noise of induction machines is generated of the rotor who has eccentricity. Given that the, that machine is equipped with a ventilator, there are two sources of noise. The influence of the fan but will not cause displacement of only a specific part of one rotation.

Given that the measured induction motor was not equipped with cooling system (fan) can be assumed, that the vibration and thus the noise are produced only by electromagnetic source and mechanical source.

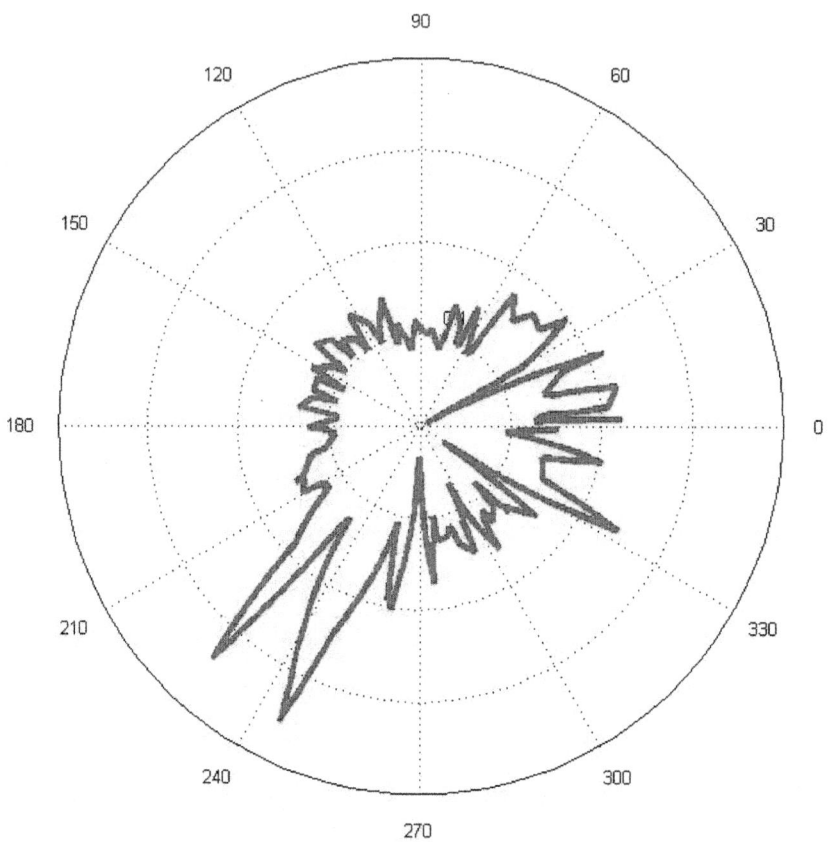

Figure 8. Noise envelope – 1 rotation.

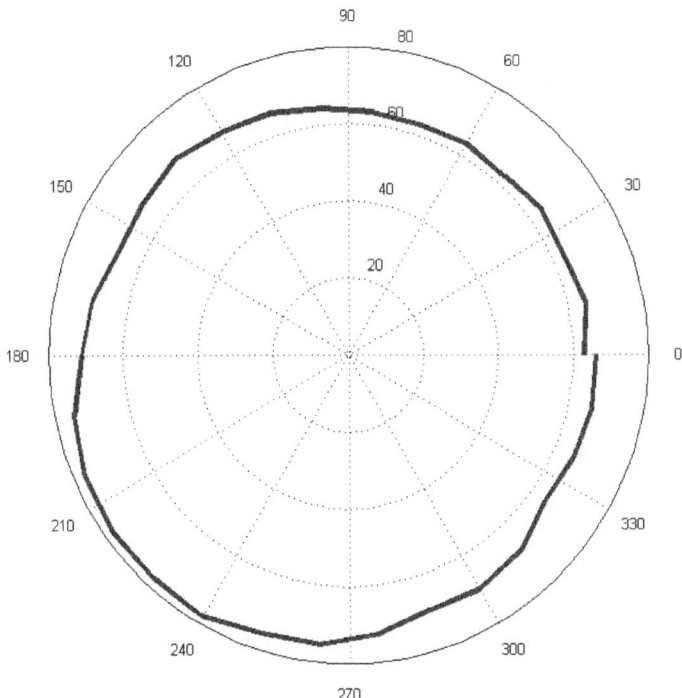

Figure 9. Rotor Eccentricity - Mechanical measurement.

Analyze of noise was performed on the one rotation of rotor. The Fig 5 shows the noise levels depending on the position of the rotor. As the graph shows it is to generate greater levels of noise in the position of the rotor from 300 to 60 degrees (about 120 degrees).

CONCLUSION

Diagnosis of induction motors is a very complex issue that has many components. One of them is the analysis of motor noise. Noise measurement asynchronous machines are the commonly used diagnostic method. This method is relatively simple. You need to be near an electrical machine quality microphone and recording equipment. Analysis itself can be the performed on specialized software, either on the spot or later in laboratory.

Subsequent analysis of the signal can then indicate whether the machine operates as required, or whether there was damage to electrical equipment.

Based on the fast Fourier analysis of noise can be determined which components of the signal are dominant. Based on knowledge of layout design of the engine is then possible to determine what is causing individual harmonics. According to the frequency it is possible to determine which there the main sources of noise are.

A major problem in measuring the noise may be interference from nearby sources. To avoid the external influence of external noise is possible only in specialized laboratories.

During measurements realized appeared possible link between noise and rotor eccentricity of electrical machinery. In the analysis of noise is dominant skew in the range of 120° in one rotation. In the same range (120°) was measured the dominant deflection of rotor eccentricity this rotor machine. Given that the machine has not a cooling system, there is not source of aerodynamic noise; there are only two possible causes of this deviation. Source of electromagnetic noise would not cause deviation only at certain rotor position, but in the whole rotation. Displacement of noise in a certain position the rotor it cannot assign too resources source of mechanical noise. This group includes vibration bearings. During the measurement was verified that the bearings are not damaged. There are not larger deviations of movement in rotation of bearing.

As a source of noise is impact of rotor eccentricity on the running of the induction motor. Unfortunately, the verification of this theory would require accurate measurement with recording of the rotor position and size of air gap. This measurement is very difficult.

ACKNOWLEDGEMENT

Research described in this paper was financed by the Ministry of Education of the Czech Republic, under project FR-TI3/073 Research and development of small electric machines; and the project of the Grant agency CR No. 102/09/1875 - – Analysis and Modeling of Low Voltage Electric Machines Parameters. The work was supported by Centre for Research and utilization of renewable energy - CZ.1.05/2.1.00/01.0014

REFERENCES

1. P Vijayraghavan, R Krishnan, Noise in electric machines: a review," Industry Applications Conference, 1998 Thirty-Third IAS Annual Meeting. The 1998 IEEE,1no., 251258vol.1, 12-15 Oct 1998doi:IAS.1998.732298
2. A. J Ellison, S. J Yang, Effects of rotor eccentricity on acoustic noise from induction machinesElectrical Engineers, Proceedings of the Institution of, 1181174184January 1971doi:piee.1971.0028

3. Hamata Václav. Hluk elektrických strojů. Praha : Academia, 1987s.
4. Mišun Vojtěch. Vibrace a hluk. 2. Brno :CERM, s.r.o., 2005s. 8-02163-060-5
5. Gieras Jacek F.; Wang, Chong; Cho Lai, Joseph. Noise of Polyphase Electric Motorss.l.] :CRC Press, 2005s. 978-0-82472-381-1
6. K. N Srinivas, R Arumugam, Analysis and characterization of switched reluctance motors: Part II. Flow, thermal, and vibration analyses",IEEE Transactions on Magnetics41413211332April 2005doi:TMAG.2004.843349. 0018-9464
7. K. N Srinivas, R Arumugam, Static and dynamic vibration analyses ofs witched reluctance motors including bearings, housing, rotor dynamics, and applied loads", IEEE Transactions on Magnetics, 40419111919July 2004doi:TMAG.2004.828034. 0018-9464
8. O. K Ersoy, A comparative eview of real and complex Fourier-related transforms" Proceedings of the IEEE, 823429447Mar 1994doi:0018-9219

Chapter 10

AUTOMOTIVE APPLICATIONS OF ACTIVE VIBRATION CONTROL

Ferdinand Svaricek[1], Tobias Fueger[1], Hans-Juergen Karkosch[2], Peter Marienfeld[2] and Christian Bohn[3]
[1]University of the German Armed Forces Munich
[2]ContiTech Vibration Control GmbH
[3]Technical University Clausthal Germany

INTRODUCTION

In recent years, commercial demand for comfortable and quiet vehicles has encouraged the industrial development of methods to accommodate a balance of performance, efficiency, and comfort levels in new automotive year models. Particularly, the noise, vibration and harshness characteristics of cars and trucks are becoming increasingly important (see, e.g., (Buchholz, 2000), (Capitani et al., 2000), (Debeaux et al., 2000), (Haverkamp, 2000), (Käsler, 2000), (Wolf & Portal, 2000), (Sano et al., 2002), (Mackay & Kenchington, 2004), (Elliott, 2008)).

Research and development activities at ContiTech and the UniBwM have focused on the transmission of engine-induced vibrations through engine and powertrain mounts into the chassis (Shoureshi et al., 1997), (Karkosch et al., 1999), (Bohn et al., 2000), (Svaricek et al., 2001), (Bohn et al., 2003), (Kowalczyk et al., 2004), (Bohn et al., 2004), (Kowalczyk & Svaricek, 2005) (Kowalczyk et al., 2006), (Karkosch & Marienfeld, 2010). Engine and powertrain mounts are usually designed according to criteria that incorporate trade-offs between vibration isolation and engine movement since the mounting system in an automotive vehicle has to fulfil the following demands:

- holding the static engine load,
- limiting engine movement due to powertrain forces and road excitations, and
- isolating the engine/transmission unit from the chassis.

Rubber and hydro mounts are the standard tool to isolate the engine and the transmission from the chassis. Rubber isolators work well (in terms of isolation) when the rubber exhibits low stiffness and little internal damping. Little damping, however, leads to a large resonance peak which can manifest itself in excessive engine movements when this resonance is excited (front end shake). These movements must be avoided in the tight engine compartments of today's cars. A low stiffness, while also giving good isolation, leads to a large static engine displacement and to a low resonance frequency (which would adversely affect the vehicle comfort and might coincide with resonance frequencies of the suspension system).

Classical mount (or suspension) design therefore tries to achieve a compromise between the conflicting requirements of acceptable damping and good isolation. It is clear that this, as well as other passive vibration control measures, are trade-off design methods in which the properties of the structure must be weighted between performance and comfort. An attractive alternative that overcomes the limitations of the purely passive approach is the use of active noise and vibration control techniques (ANC/AVC). The basic idea of ANC and AVC is to superimpose the unwanted noise or vibration signals with a cancelling signal of exactly the same magnitude and a phase difference of 180° (i.e. the "anti-noise" principle of Lueg (Lueg, 1933)). In the case of ANC, this cancelling signal is generated through loudspeakers, whereas for AVC, force actuators such as inertia-mass shakers are used. Various authors have addressed the application of ANC and AVC systems to reduce noise and vibrations in automotive applications (Adachi & Sano, 1996), (Adachi & Sano, 1998), (Ahmadian & Jeric, 1999), (Bao et al., 1991), (Doppenberg et al., 2000), (Dehandschutter & Sas, 1998), (Fursdon et al., 2000), (Lecce et al., 1995), (Necati et al., 2000), (Pricken, 2000), (Riley & Bodie, 1996), (Sas & Dehandschutter, 1999), (Shoureshi et al., 1995), (Shoureshi et al, 1997), (Shoureshi & Knurek, 1996), (Sano et al., 2002), (Swanson, 1993). ContiTech has implemented prototypes of AVC systems in various test vehicles and demonstrated that significant reductions in noise and vibration levels are achievable (Shoureshi et al., 1997), (Karkosch et al., 1999), (Bohn et al., 2000), (Svaricek et al., 2001), (Bohn et al., 2003), (Kowalczyk et al., 2004), (Bohn et al., 2004), (Kowalczyk & Svaricek, 2005) (Kowalczyk et al., 2006), (Karkosch & Marienfeld, 2010). Honda has developed a series-production ANC/AVC system to reduce noise and vibration due to cylinder cutoff in combination with the engine RPM as reference signal (Inoue et al., 2004), (Matsuoka et al., 2004). A recent overview of such series-production AVC systems can be found in (Marienfeld, 2008).

Most of these approaches rely on feedforward control strategies (either pure feedforward or combined with feedback). The feedforward signal is either taken from an additional sensor (usually an accelerometer in active vibration control) or generated artificially from measurements of the fundamental disturbance frequency (Kuo & Morgan, 1996), (Hansen & Snyder, 1997), (Clark et al., 1998), (Elliot, 2001). Contrary to the major fields of application for active noise and vibration control (military and aircraft), the automotive sector is extremely sensitive to the costs of the overall system. It is therefore desirable to use an approach that requires only one sensor. Also, most approaches rely on adaptive control strategies such as the filtered-x LMS algorithm (Kuo & Morgan, 1996), (Hansen & Snyder, 1997), (Clark et al., 1998), (Elliot, 2001). This seems necessary as the characteristics of the disturbance acting upon the system are time varying. In automotive applications, for example, the fundamental frequency (engine firing frequency, which is half the engine speed in four-stroke engines) varies from 7 Hz at idle to 50 Hz at 6000 rpm. The adaptive approach will adjust the disturbance attenuation of the control system to the frequency content of the disturbance. Whereas this works well in many applications (see the references given above), some critical issues such as convergence speed, tuning of the step size in the adaptive algorithm and stability remain. Discussions between the authors and potential customers (automobile manufactures) have indicated that particularly the issues of convergence speed, tracking performance (this is related to the attenuation capability of the algorithm during changes in engine speed such as fast acceleration) and stability are crucial. A non-adaptive algorithm might have the benefit of a higher customer acceptance. Another advantage of a non-adaptive algorithm is that the behavior of the closed-loop system can be analysed independent of the input signals. In an adaptive algorithm, the optimal controller depends on the external signals that act upon the system; thus, it is very difficult to analyse the performance off-line.

Both kind of algorithms have been implemented in an active control system for cancellation of engine-induced vibrations in several test vehicles. The remainder will present an overview of ANC/AVC system components, control algorithms, as well as obtained experimental results.

SYSTEM DESCRIPTION

A schematic representation of an AVC system in a vehicle is shown in Figure 1. The disturbance force originating from the engine and transmitted into the chassis through the engine mounts is actively cancelled by an actuator force of the same magnitude but of opposite sign.

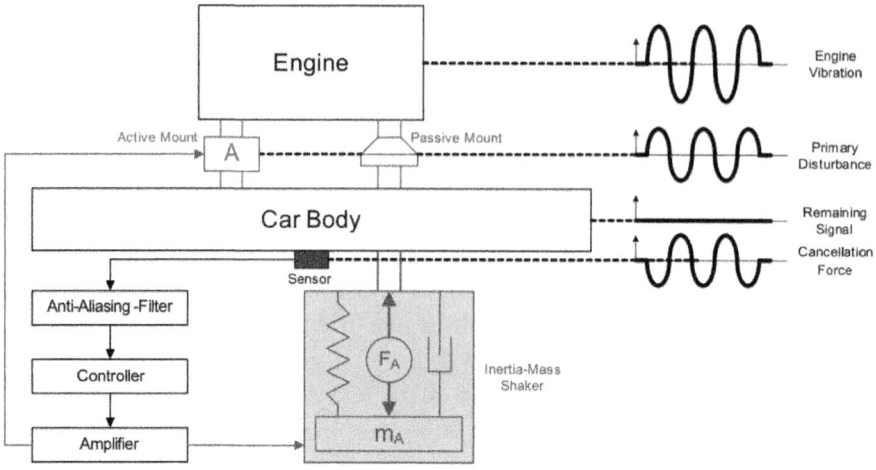

Figure. 1. Schematic representation of an AVC system with an active mount (red) or an inertiamass shaker (green).

The basic components of the system include actuators, sensors and an electronic control unit (ECU). The electronic hardware consists of an amplifier and filter unit that contains the power amplifier and the anti-aliasing filter for the sensor signal. Figure 1 presents two alternative principles, the inertial mass actuator (green) and active mount with integrated actuator (red), whereby the ECU, sensors and the actuator's basic components can be identically used. The two principles are similar in how they function, forces are fed into the system in targeted fashion so that the resulting dynamic forces at the base of the mount (attachment point) are reduced. In this example, attachment point acceleration is measured and supplied to the controller. The countersignal calculated in the control unit powers the actuator via power amplifiers. Ideally, the superimposed forces cancel out one another so that no annoying engine vibration is disseminated via the chassis.

Generally, there are two possible ways of active vibration cancellation, the Inertial mass shakers, attached at suitable points, cancel out the disturbing vibration by a force signal of opposite phase. On the other hand, active engine mounts compensate the displacement between engine mount and the car body. Hereby the car body is kept free from the vibration forces emitted from the engine. With regard to the specifications of the AVC system the suitable system configuration has to be chosen. In (Hartwig et al., 2000), (Karkosch & Marienfeld, 2010) the electrodynamic and the electromagnetic actuator principle and the two system configurations are compared. Figure 2 shows the electrodynamic and electromagnetic actuator principles.

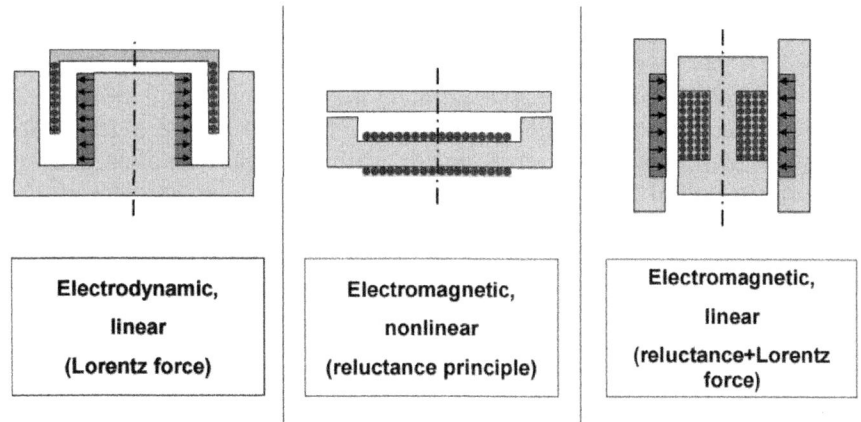

Figure. 2. Electrodynamic and electromagnetic actuator principles.

An electromagnetic actuator has the benefit to an electrodynamic actuator in higher actuator force at decreased magnet and design volume such as more cost efficient. On the other side, the electrodynamic actuator principle has the advantage in a simple design of the iron core and the absence of magnetic forces lateral to the deflection direction. A comparison between an active absorber and an active hydromount configuration is shown in Figure 3.

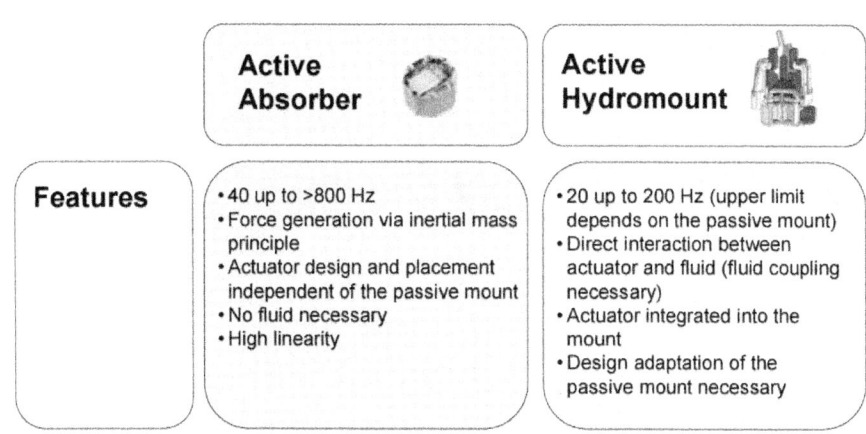

Figure. 3. Active engine mount system configurations – principle comparison.

CONTROL SYSTEM DESIGN

The problem of active control of noise and vibrations has been a subject of much research in recent years. For an overview see e.g. (Kuo & Morgan, 1996), (Hansen & Snyder, 1997), (Clark et al., 1998), (Elliot, 2001) and the references therein. The main part of the published literature makes use of adaptive feedforward structures. Adaptive feedback compensation (Aström & Wittenmark, 1995), in which the feedback law depends explicitly upon the error sensor output has found little application in the active noise and vibration control field. Feedforward control provides the ability to handle a great variety of disturbance signals, from pure tone to a fully random excitation. However, the performance of feedforward control algorithms can be degraded if disturbances are not measurable in advance (e.g. road or wind noise) or the transmission path characteristics change rapidly. Contrarily, a feedback controller can be designed to be less sensitive to system perturbations. Robustness and performance, however, are conflicting design requirements. To achieve a good attenuation of the vibrations the cancellation wave has to be very accurate, typically within ±5 degrees in phase and ±0.5 dB in amplitude.

FxLMS Approach

The FxLMS algorithm has been originally proposed in (Morgan, 1980) and is described in detail in (Kuo & Morgan, 1996). The basic idea is to use the feedforward structure shown in Figure 4. The transfer path between the disturbance source and the error sensor is called primary path. The secondary path is the transfer path between the output of the controller and the error sensor. The aim in the control loop is to minimize the output signal (error signal).

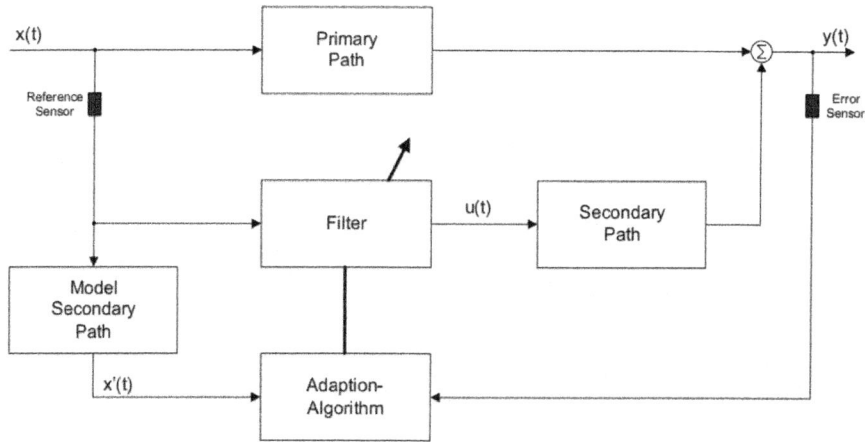

Figure. 4. Block diagram of FxLMS algorithm.

The adaptive filter has to approximate the dynamics of the primary path and the inverse dynamics of the secondary path. For the on-line adaptation of a FIR-filter (finite-impulseresponse filter), two signals are used: error signal and reference signal filtered with the model of the secondary path (filtered-x).

The discrete-time transfer function of a FIR-filter has the form

$$F(z) = \frac{U(z)}{X(z)} = \frac{w_0 z^m + w_1 z^{m-1} + \ldots + w_m}{z^m}, \quad (1)$$

whereas the filter coefficients wi, i=1,...,m can be represented as a vector:

$$w(k) = \begin{bmatrix} w_0(k) & w_1(k) & \ldots & w_m(k) \end{bmatrix}^T \quad (2)$$

The adaptation of the filter weights w_i is performed through the well-known LMS (least mean square) algorithm originally proposed in (Widrow & Hof, 1960). A performance index J is built from the sum of squares of the sampled error signal:

$$J = \frac{1}{N} \sum_{i=1}^{N} y^2(i) \quad (3)$$

This performance function depends on the filter coefficients and can be described through a hyperparaboloid as shown in Figure 5. The optimal values for the adaptive filter coefficients are located in the deepest point of the performance surface. The LMS-algorithm is searching on-line for the coordinates of the deepest point. The control signal is generated as the output of the adaptive filter.

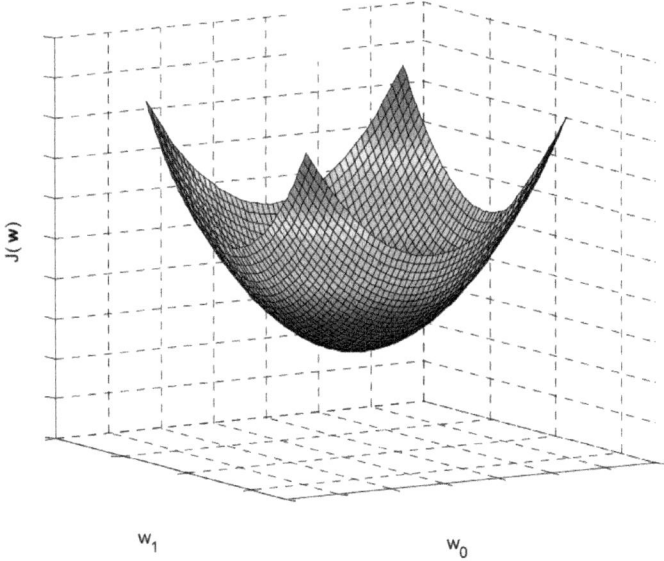

Figure. 5. Example of a performance surface for a two-weight system.

Disturbance Observer Approach

This method is based on state observer and state feedback and has been proposed in (Bohn et al., 2003), (Kowalczyk et al., 2004), (Bohn et al., 2004), (Kowalczyk & Svaricek, 2005). It is assumed that the disturbance enters at the input of the plant S, see Figure 6.

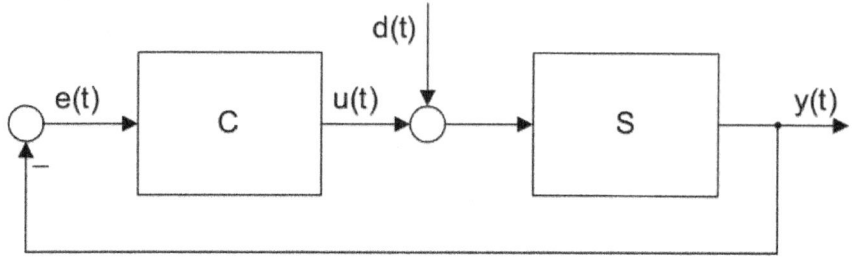

Figure. 6. Control loop with a plant S and a controller C.

The disturbance is modelled as a sum of a finite number of sine signals, which are harmonically related:

$$d(t) = \sum_{i=1}^{N} A_i \sin(2\pi f_i t + \varphi_i).$$
(4)

This disturbance is time-varying and needs frequency measurements to be fed into the model. The disturbance attenuation is achieved through producing an estimate of the disturbance d and using this estimate, with a sign reversal, as a control signal u. To generate the estimate, a disturbance observer is used. The observer is designed off-line assuming time-invariance and investigating the property of robustness over a certain frequency region for a single observer. Later on, a gain-scheduling is implemented to cover the whole frequency region of interest by a stable observer. This provides a non-adaptive approach, where the frequency is used as a scheduling variable. The transfer function of the controller C has infinite gain at the frequencies included in the disturbance model. The controller poles show up as zeros in the closed-loop transfer function. Figure 7 shows the frequency response magnitude of the sensitivity function 1/(1+CS).

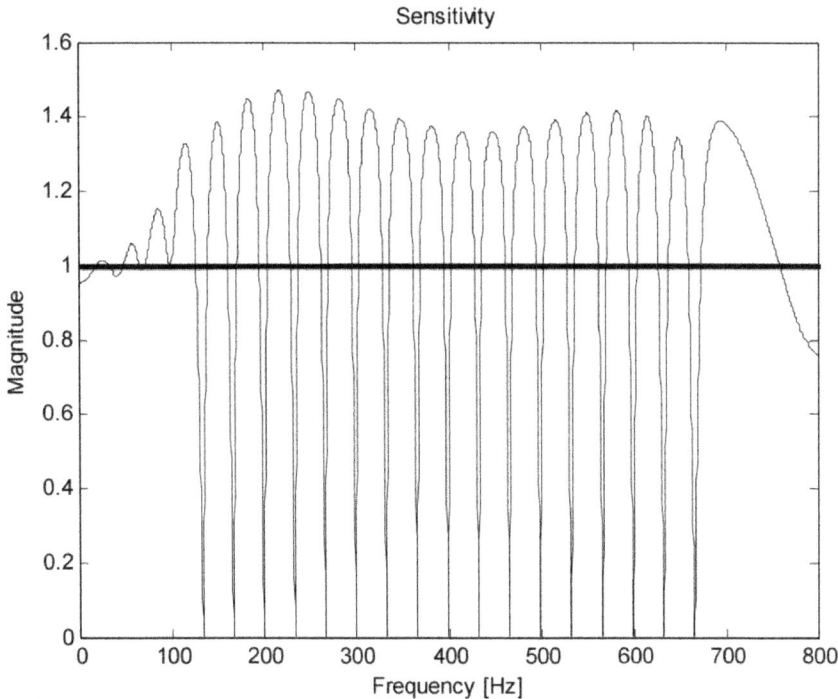

Figure. 7. Frequency response magnitude of the sensitivity function.

It can be seen that the magnitude of the sensitivity function is zero for the frequencies specified in the disturbance model, which corresponds to complete disturbance cancellation. The improvement of the disturbance attenuation for these frequencies leads to some disturbance amplification between these frequencies. This effect is in accordance with Bode's well-known sensitivity integral theorem and is called waterbed effect (Hong & Bernstein, 1998). For more details on this algorithm, see (Bohn et al., 2004). Finally, both approaches can be combined to give a two-degree-of-freedom control structure, which is referred to as a hybrid approach in the ANC/AVC literature (Shoureshi & Knurek, 1996), (Hansen & Snyder, 1997). The implementation of all control algorithms is usually done on digital signal processing hardware.

Due to a large number of influence parameters, no definite statements can be made with regard to which control scheme will give a better performance. Rather, control strategies have to be chosen with regard to the characteristics of the vibration problem to be addressed, such as available sensor signals (e.g., costs associated with additional feedforward sensors, possible use of existing sensors), Type of excitation (periodic, e.g. engine vibrations, or stochastic, e.g. road excitations), Frequency range of interest (e.g. 25 – 30 Hz for idling speed or 25 – 300 Hz for the whole engine speed range), Spectral characteristics of excitation (narrowband, e.g. distinct frequencies, or broadband; e.g. fixed/ varying frequencies).

The decision for one particular control strategy and the determination of suitable controller settings is a very important step in the development of ANC/AVC schemes. Therefore, simulation studies and real-time experiments on vehicles are carried out to identify a suitable strategy for a given noise and vibration problem. For the real-time experiments, the control strategies, together with auxiliary function such as signal conditioning and monitoring routines, are implemented on a rapid prototyping system.

EXPERIMENTAL RESULTS

In the last years, several vehicles — with different problems — have been equipped with active absorber systems to attenuate the transmission of the engine vibrations into the vehicle cabin. As mentioned earlier, the control algorithms have to be chosen with regard to the particular problem of the considered vehicle. ContiTech has equipped a test vehicle with an AVC system with inertia-mass shaker attached on the transmission cross-member. Figure 8 shows the location of the system components on the transmission cross-member in the test vehicle.

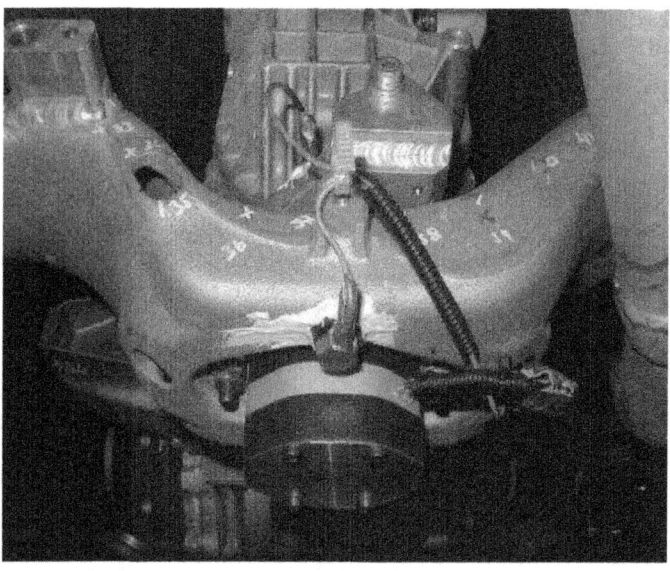

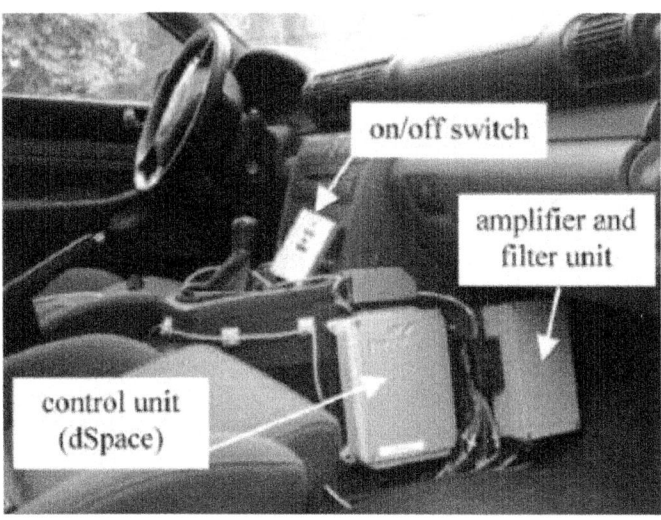

Figure. 8. Location of the AVC components in the test vehicle.

The control algorithm is implemented on a rapid prototyping unit, the dSPACE MicroAutoBox. The electronic hardware consists of an amplifier and filter unit that contains the power amplifier and the anti-aliasing filter for the sensor signal, and the electronic control unit. A remote control on/off switch is used to turn the control algorithm on and off during vehicle tests (Kowalczyk et al., 2006). In control engineering terms, the transfer function from the amplifier input to the (filtered) sensor output is the transfer function of the plant to be controlled (assuming linearity and time invariance). In accordance with the active noise and vibration control literature (Kuo & Morgan, 1996), (Hansen & Snyder, 1997), (Clark et al., 1998) this is called the secondary path S. To design a control algorithm, a model for the secondary path is required. Quite often models for vibration control systems are derived from physical principles (Preumont, 1997), from finite-element models or through experimental techniques such as modal analysis (Heylen et al., 1997). Physical principles are mostly applied to fairly simple mechanical structures such as beams or plates for which analytical solutions can be found. Finiteelement models or models derived from modal analysis will give a model of the structure only, that is, without the dynamics of the electrical and electromechanical components (amplifier, actuator, sensor). The approach taken here is to excite the system with a test signal and record the response. Any of the discrete-time black-box system identification techniques (such as the least squares approach for equation-error models) can then be used to identify a model (Ljung & Söderström, 1983). Figure 9 shows the amplitude and phase responses of an identified system transfer function. The amplitude response would be dimensionless, since it corresponds to the output voltage, i.e. the filtered sensor signal, over the input voltage of the amplifier. However, for interpretability, the output signal has been scaled to acceleration (m/s², using the sensor sensitivity) and the input signal to current (A, using the amplifier gain). Such models are used for the subsequent controller design and for simulation studies.

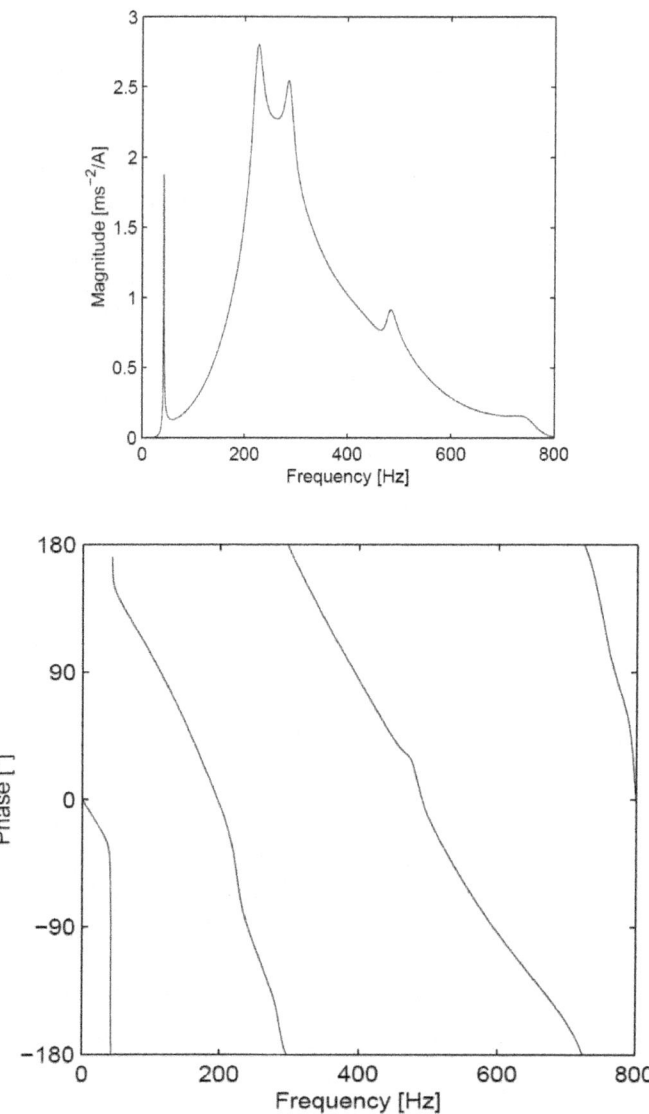

Figure. 9. Amplitude and phase plots of an identified system transfer function (actuator current to filtered sensor output; the first peak corresponds to the resonance of the inertia-mass actuator).

For instance, the stationary behavior of the controlled system is of interest when the comfort under idling speed conditions should be improved. A typical real-time result for such a problem is given in Figure 10. Here, a comparison of the error signals (measured accelerations at the frame) is shown for control off and on. It can be seen that the engine orders 2, 4 and 6 are predominant at idling speed without active control. However, a significant reduction (up to 37 dB) of these engine orders can be achieved by using an AVC system.

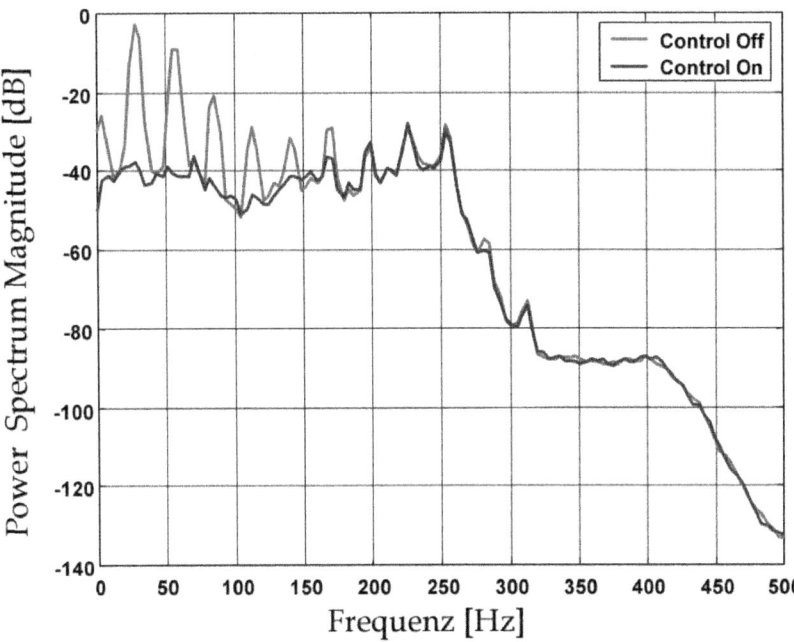

Figure. 10. Power spectrum of the measured frame vibrations at idle.

In other applications, the active system should work over a wide engine speed range. For such applications the tracking behavior of the active system must be considered. Figure 11 gives an impression of the dynamic behavior of the adaptive FxLMS algorithm. To illustrate the adaptation of the controller, the decrease in the measured frame vibrations after switching on the control algorithm at t=1[s] is shown.

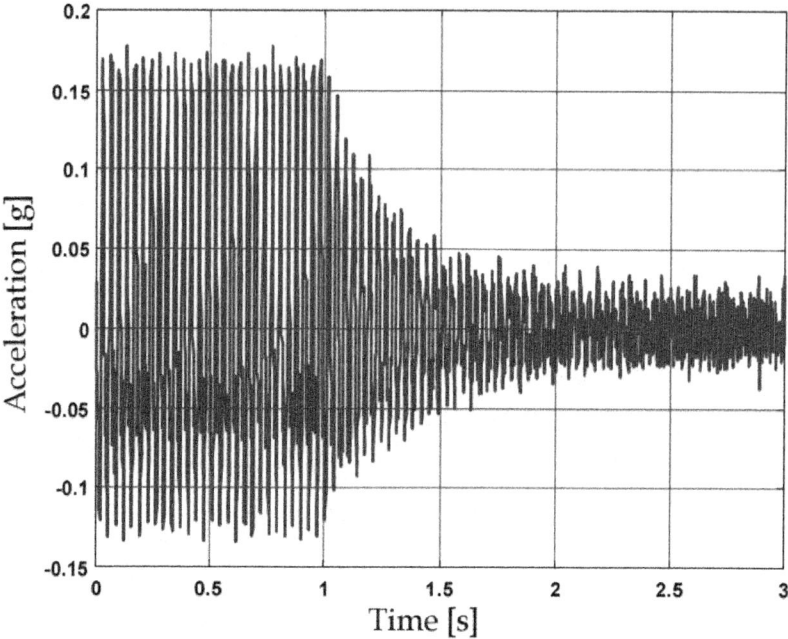

Figure. 11. Adaptation behavior of the FxLMS algorithm.

It is well–known that parts of the transmitted vibration energy through the mounts pass through the chassis and emanate in the vehicle passenger compartment in the form of structure–borne noise. Figure 12 shows an order analysis of a sound pressure level measurement at the passenger's left ear of a test vehicle that has acoustic problems in the frequency range between 200 and 300 Hz.

Here a lot of engine orders (2.5, 3, 3.5, ...) are visible since the transmission mount is the major path for this engine–induced noise. The improvement with control on is shown in Figure 13. The sound pressure level measurement at the passenger's left ear points out a significant reduction in sound for frequencies higher than 120 [Hz].

Due to the fact that the measured vibrations at the transmission are well correlated to the cross member vibrations a classical FxLMS algorithm has been chosen for this application. An impressive reduction of the sound pressure level, achieved by the small (weight about 0.6 kg) active absorber at the transmission mount, can be registered in Figure 13. The remaining 2nd order line is a result of the vibrations that are still transmitted through the two front engine mounts.

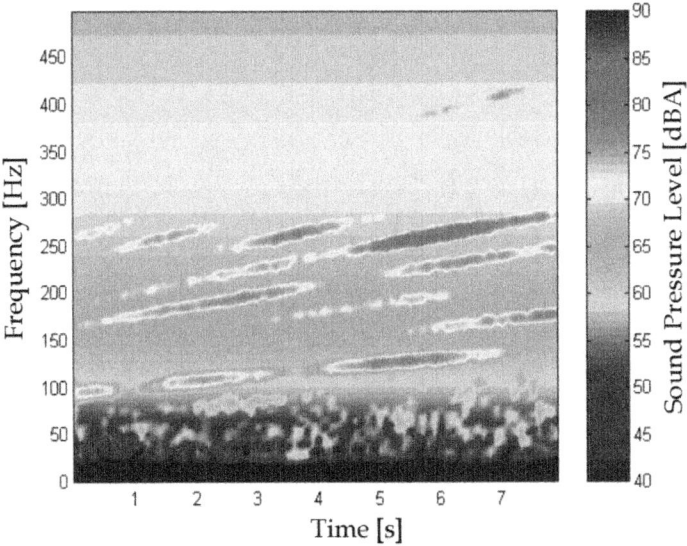

Figure. 12. Order analysis of sound pressure level (passenger's left ear) of a road test (acceleration from 1800 to 4500 rpm, full throttle, 3rd gear, control off).

The active absorber system has not only a great impact on the interior noise of the vehicle but also on vibrations at comfort relevant points. Such an interior comfort improvement for the passengers can be observed from a control on/off comparison of the power spectrum of the measured acceleration signal at the steering wheel, see Figure 14.

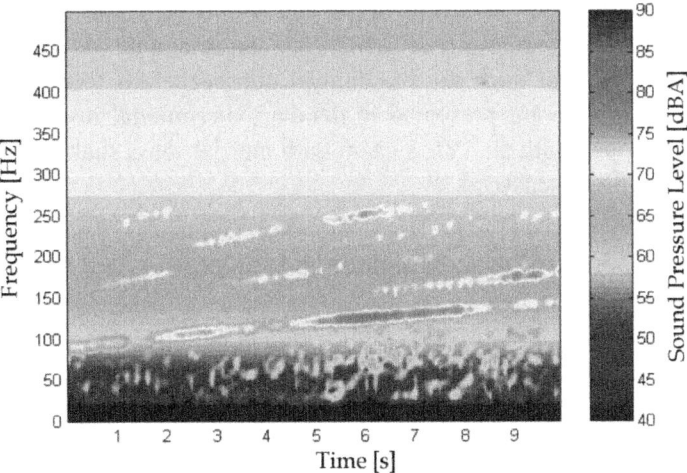

Figure. 13. Order analysis of sound pressure level (passenger's left ear) of a road test (acceleration from 1800 to 4500 rpm, full throttle, 3rd gear, control on).

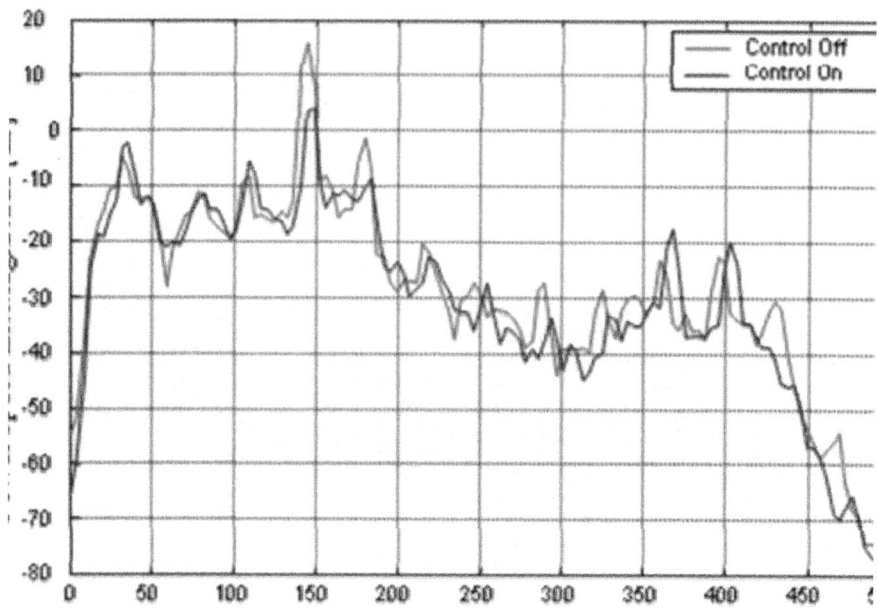

Figure. 14. Power spectrum comparison of the measured steering wheel acceleration for constant drives with 4400 RPM.

CONCLUSION

This chapter has given an overview of recent research and development activities in the field of active noise and vibration control in automotive applications. The design of an ANC/AVC system with its components is described in general such as two control approaches, a feedforward and a feedback approach, are presented in detail. Experimental results from a test vehicle, equipped with an AVC system with inertial-mass shaker and a dSpace MicroBox, were discussed.

Recent advances in NVH (Noise Vibration Harshness) design and analysis tools, development of low cost digital signal processors, and adaptive control theory, have made active vibro–acoustic systems a viable and economically feasible solution for low frequency problems in automotive vehicles.

Further experimental results and a comparison of the presented control approaches can be found in (Kowalczyk et al., 2004) and (Kowalczyk & Svaricek, 2005).

REFERENCES

1. Adachi, S. & Sano, H. (1996). Application of a two-degree-of-freedom type active noise control using IMC to road noise inside automobiles. Proceedings of the 35th IEEE Conference on Decision and Control, pp. 2794-2795, Kobe
2. Adachi, S. & Sano, H. (1998). Active noise control system for automobiles based on adaptive and robust control. Proceedings of the 1998 IEEE International Conference on Control Applications, pp. 1125-1128, Trieste
3. Ahmadian, M. & Jeric, K.M. (1999). The application of piezoceramics for reducing noise and vibration in vehicle structures. SAE Technical Paper 1999-01-2868. Proceedings of the International Off-Highway and Powerplant Congress and Exposition, Indianapolis
4. Aström, K.J. & Wittenmark, B. (1995). Adaptive Control, Addison–Wesley, Reading
5. Bao, C.; Sas, P. & Van Brussel, H. (1991). Active control of engine-induced noise inside cars. Proceedings of the International Conference on Noise Control Inter-noise 91, pp. 525-528, Sydney
6. Bohn, C.; Karkosch, H.-J.; Marienfeld, P.M. & Svaricek, F. (2000). Automotive applications of rapid prototyping for active vibration control. Proceedings of the 3rd IFAC Workshop Advances in Automotive Control, pp. 191-196, Karlsruhe, Germany
7. Bohn, C.; Cortabarria, A.; Härtel, V. & Kowalczyk, K. (2003). Disturbance-observer-based active control of engine-induced vibrations in automotive vehicles. Proceedings of the 10th Annual International Symposium on Smart Structures and Materials. Paper 50, pp. 49-68, San Diego, USA
8. Bohn, C.; Cortabarria, A.; Härtel, V. & Kowalczyk, K. (2004). Active control of engineinduced vibrations in automotive vehicles using disturbance observer gain scheduling. Control Engineering Practice 12, 1029-1039.
9. Buchholz, K. (2000). Good vibrations. Automotive Engineering, 108, (August 2000) 85-89
10. Capitani; Citti, R.P.; Delogu, M.; Mascellini, R. & Pilo, L. (2000). Experimental validation of a driveline numerical model for the study of vibrational comfort of a vehicle. Proceedings of the 33rd ISATA Electric, Hybrid, Fuel Cell and Alternative Vehicles/Powertrain Technology, pp. 521-530, Dublin
11. Clark, R.L.; Saunders, W.R. & Gibbs, G.P. (1998). Adaptive Structures:

Dynamics and Control, John Wiley and Sons, New York
12. Debeaux, E.; Claessens, M. & Hu, X. (2000). An analytical-experimental method for analysing the low-frequency interior acoustics of a passenger car. Proceedings of the 2000 International Conference on Noise and Vibration Engineering ISMA 25, pp. 1331- 1338, Leuven
13. Dehandschutter, W. & Sas, P. (1998). Active control of structure-borne road noise using vibration actuators. Journal of Vibration and Acoustics 120:517-523 Doppenberg, E.J.J.; Berkhoff, A.P. & van Overbeek, M. (2000). Smart materials and active noise and vibration control in vehicles. Proceedings of the 3rd IFAC Workshop Advances in Automotive Control, pp. 205-214, Karlsruhe, Germany
14. Elliott, S.J. (2001). Signal Processing for Active Control, San Diego, Academic Press
15. Elliott, S.J. (2008). A Review of Active Noise and Vibration Control in Road Vehicles. ISVR Technical Memorandum No. 981, University of Southhampton
16. Fursdon, P.M.T.; Harrison, A.J. & Stoten, D.P. (2000). The design and development of a selftuning active engine mount. Proceedings of the European Conference on Noise and Vibration 2000, pp. 21-32, London
17. Hansen, C.H. & Snyder, S.D. (1997). Active Control of Noise and Vibration, E & FN, London
18. Hartwig, C.; Haase, M.; Hofmann, M. & Karkosch, H.-J. (2000). Electromagnetic actuators for active engine vibration cancellation. Proceedings of the 7th International Conference on New Actuators ACTUATOR 2000, Bremen, June 2000
19. Haverkamp, M. (2000). Solving vehicle noise problems by analysis of the transmitted sound energy. Proceedings of the 2000 International Conference on Noise and Vibration Engineering ISMA25, pp.1339-1346, Leuven
20. Heylen, W.; Lammens, S. & Sas, P. (1997). Modal Analysis Theory and Testing, Katholieke Universiteit Leuven Departement Werktuigkunde, Leuven
21. Hong, J. & Bernstein, D.S. (1998). Bode integral constraints, colocation, and spillover in active noise and vibration control. IEEE Transactions on Control Systems Technology 6, 111-120
22. Inoue, T. ; Takahashi, A.; Sano, H.; Onishi, M. & Nakamura, Y. (2004). NV Countermeasure Technology for a Cylinder-On-Demand Engine-Development of Active Booming Noise Control System Applying

Adaptive Notch Filter. SAE-Paper 2004-01-0411. Noise and Vibration 2004 SP-1867. 131-138

23. Käsler, R. (2000). Development trends and vibro-acoustic layout criteria for powertrain mounting systems. Proceedings of the International Congress Engine & Environment 2000, pp. 155-172, Graz

24. Karkosch, H.-J.; Svaricek, F.; Shoureshi, R.A. & Vance, J.L. (1999). Automotive applications of active vibration control. Proceedings of the European Control Conference, Karlsruhe

25. Karkosch, H.-J. & Marienfeld, P.M. (2010). Use of Active Engine Mounts to Optimize Comfort in Cars with Innovative Drives. Proceedings of the 12th International Conference on New Actuators ACTUATOR 2010, Bremen, June 2010

26. Kowalczyk, K.; Svaricek, F. & Bohn, C. (2004). Disturbance-observer-based active control of transmission-induced vibrations. Proceedings IFAC Symposium Advances in Automotive Control, pp. 78-83, Salerno, Italy

27. Kowalczyk, K. & Svaricek, F. (2005). Experimental Robustness of FXLMS and DisturbanceObserver Algorithms for Active Vibaration Control in Automotive Applications. In Proceedings of the 16th IFAC World Congress, Prag

28. Kowalczyk, K.; Karkosch, H.-J.; Marienfeld, P.M. & Svaricek, F. (2006). Rapid Control Prototyping of Active Vibration Control Systems in Automotive Applications. Proceedings of the 2006 IEEE International Conference on Computer Aided Control Systems Design, Munich, pp. 2677-2682

29. Kuo, S.M. & Morgan, D.M. (1996). Active Noise Control Systems, John Wiley and Sons, New York

30. Lecce, L.; Franco, F.; Maja, B.; Montouri, G. & Zandonella-Necca, D. (1995). Vibration active control inside a car by using piezo actuators and sensors. 28th International Symposium on Automotive Technology and Automation. Proceedings for the Dedicated Conference on Mechatronics – Efficient Computer Support for Engineering, Manufacturing, Testing and Reliability. Croydon, pp. 423-432, UK

31. Ljung, L. & Söderström, T. (1983). Theory and Practice of Recursive Identification, MIT Press, Cambridge

32. Lueg, P. (1933). Process of silencing sound oscillations. US Patent No 2,043,416. Filed: March 8, 1934. Patented: June 6, 1936. Priority (Germany): January 1933

33. Mackay, A.C. and Kenchington, S. (2004). Active control of noise and vibration – A review of automotive applications. Proceedings ACTIVE 2004, Williamsburg
34. Marienfeld, P. (2008). Übersicht über den Serieneinsatz mechatronischer Systeme im Bereich der Aggregatelagerung. Tagung „Geräusch- und Schwingungskomfort von Kraftfahrzeugen", Haus der Technik, Munich
35. Matsuoka, H. ; Mikasa, T. & Nemoto, H. (2004). NV Countermeasure Technology for a Cylinder-On-Demand Engine- Development of Active Control Engine Mount. SAEPaper 2004-01-0413. Noise and Vibration 2004 SP-1867
36. Morgan, D.R. (1980). An Analysis of Multiple Correlation Cancellation Loops with a Filter in the Auxiliary Path. IEEE Trans. Acoust., Speech, Signal Processing 28, 454-467
37. Necati, G.A.; Doppenberg, E.J.J. & Antila, M. (2000). Noise radiation reduction of a car dash panel. Proceedings of the 2000 International Conference on Noise and Vibration Engineering ISMA25, pp. 855-862, Leuven
38. Preumont, A. (1997). Vibration Control of Active Structures, Kluwer Academic Publishers, Dordrecht, The Netherlands
39. Pricken, F. (2000). Active noise cancellation in future air intake systems. SAE-Paper 2000-01- 0026. Powertrain Systems NVH. SAE Special Publication SP-1515. 1-6
40. Riley, B. & Bodie, M. (1996). An adaptive strategy for vehicle vibration and noise cancellation. Proceedings of the IEEE 1996 National Aerospace and Electronics Conference NAECON 1996, pp. 836-843, Dayton
41. Sano, H.; Yamashita, T. & Nakamura, M. (2002). Recent application of active noise and vibration control to automobiles. Proceedings ACTIVE 2002, pp. 29-42, Southampton, UK
42. Sas, P. & Dehandschutter, W. (1999). Active structural and acoustic control of structureborne road noise in a passenger car. Noise & Vibration Worldwide 30, 17-27
43. Shoureshi, R.A.; Alves, G.; Knurek, T.; Novotry, D.; Ogundipe, L. & Wheeler, M. (1995). Mechatronically-based vibration and noise control in automotive systems. 28th International Symposium on Automotive Technology and Automation. Proceedings for the Dedicated Conference on Mechatronics – Efficient Computer Support for Engineering, Manufacturing, Testing and Reliability, pp. 691-698, Croydon, UK
44. Shoureshi, R. & Knurek, T. (1996). Automotive applications of a hybrid

active noise and vibration control. IEEE Control Systems Magazine 16, 72-78
45. Shoureshi, R.A.; Gasser, R. & Vance, J.L. (1997). Automotive applications of a hybrid active noise and vibration control. Proceedings of the IEEE International Symposium on Industrial Electronics, pp. 1071-1076, Guimaraes, Portugal
46. Shoureshi, R.A.; Vance, J. L.; Ogundipe, L.; Schaaf, K.; Eberhard, G. & Karkosch, H.-J. (1997). Active vibro-acoustic control in automotive vehicles. Proceedings of the 1997 Noise and Vibration Conference, pp. 131-136, Traverse City, MI
47. Svaricek, F.; Bohn, C.; Karkosch, H.-J. & Härtel, V. (2001). Aktive Schwingungskompensation im Kfz aus regelungstechnischer Sicht. at – Automatisierungstechnik 49, 249-259. (Active vibration cancellation in automotive vehicles from a control engineering point of view, in German)
48. Swanson, D.A. (1993). Active engine mounts for vehicles. SAE Technical Paper 932432. Proceedings of the 1993 International Off-Highway and Powerplant Congress and Exposition, Milwaukee
49. Widrow, B. & Hof, M.E. (1960). Adaptive Switching Circuits. IRE WESCON Conv. Rec. 96- 104
50. Wolf, A. & Portal, E. (2000). Requirements to noise reduction concepts and parts in future engine compartments. SAE-Paper 2000-01-0027. Powertrain Systems NVH. SAE Special Publication SP-1515, pp. 7-12.

Chapter 11

PROGRESS AND RECENT TRENDS IN THE TORSIONAL VIBRATION OF INTERNAL COMBUSTION ENGINE

Liang Xingyu, Shu Gequn, Dong Lihui, Wang Bin and Yang Kang

State Key Laboratory of Engines, Tianjin University, 300072 P. R. China

INTRODUCTION

With modern machinery industry developing, the application of internal combustion engine is getting wider and research direction is towards high-power, high speed and strong loads. So the issue of torsional vibration of the engine is becoming more prominent. All kinds of work conditions of the engine may have great impacts on the shafting, leading to all sorts of torsional vibration and resonance, and many accidents which lead to much detriment have occurred at home and abroad due to torsional vibration. As the problem of torsional vibration of the engine is becoming more and more prominent, broad research is made both at home and abroad. This article mainly refers to the literatures on torsional vibration issue published in recent years, summarizes on the modeling of torsional vibration, corresponding analysis methods, appropriate measures and torsional vibration control, and points out the problems to be solved in the study and some new research directions.

MODELING OF ENGINE CRANKSHAFT

Engine Crankshaft Modeling Method

Crankshaft is the main component of internal combustion engine. Shaft vibration is one of the most important factors affecting engine operation safety. Crankshaft modeling is the base of crankshaft torsional vibration analysis, whose accuracy and simple practical applicability will greatly improve the efficiency and credibility of research results. At present, there are 3 kinds of most basic shaft models used in analyzing torsional vibration: the first type is simple mass - spring model, the second is continuous mass model, and the third is multi-segment concentrated mass model.

Simple Mass - Spring Model

Simple mass - spring model is the earliest mechanics model in the calculation of shaft vibration [1-6], which was also called lumped parameter model in some literatures. It disperses crankshaft onto the disk with concentration of inertia moment, elastic axis without mass, internal damping and external damping, as shown in figure 1. Each disk rotational inertia includes: the rotational inertia of the crank, the equivalent rotational inertia of connecting rod and piston, transmission system, shock absorber, the rotational inertia of the flywheel, etc.

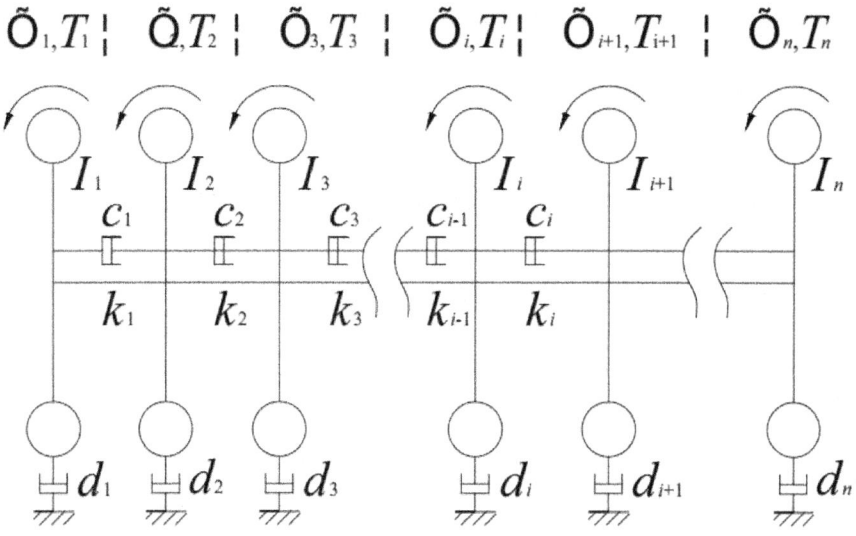

Figure. 1. Simple mass - spring model schematic diagram.

This model has certain precision for lower frequency of torsional vibration modal and clear physical concept. It's simple to use and easy to calculate. But since this model is simplified, when precise calculation of the crankshaft is required, its precision is limited. This model is established completely for rigid shaft and rotation parts, so it cannot simulate the actual shaft.

Continuous Mass Model

Continuous mass model is based on continuum theory, regarding shaft as elastomer, established in finite element method. It's also called distributed mass model in some literatures [7-8]. It adopts finite element method in general, dissecting the crankshaft entities directly into finite element calculation model of division. Hence, the mass of the shaft is distributed continuously along the shaft, closer to practice than that of simple mass - spring model. Partial differential equations can be used in this model, which can accurately calculate low frequency and vibration model of the shaft, as well as high frequency and vibration model, solve by numerical method, and also can calculate arbitrary section stress conveniently. But the model is complex and with low speed to calculate, and is easy to cause greater accumulative error. It is more difficult to use this model in system simulation and design. Due to the method of forced vibration calculation, it is hard to realize, thus it's mainly used in the calculation of free vibration.

Recently, two consecutive quality models also have derived from this model: framework model and multi-diameter model.

Framework model is a model, in which, circular cross section straight beam represents main journal and crank pin, and variable cross-section rectangular beam represents crank arm and counterbalance in finite element analysis [9]. For these analyses, circular cross section beam also can represent main journal and crank pin, but the crank arm and counterbalance should be treated as simple rectangular beam. Model schematic diagram is shown in figure 2. In framework model, different structural parts of the crankshaft are substituted by the continuous entities with regular shape, and the original basic shapes of crankshaft are kept. Thus this model has higher precision to analyze the crankshaft vibration.

Multi-diameter model is a model used in elastic wave propagation theory solving torsional vibration of internal combustion engine [10-12]. Assign piston-rod additional mass to two crank arms and simplify a unit crankcase into a group of concentric multi-diameter. Model schematic diagram is shown in figure 3. Because the model has continuous mass distribution, the effect of distribution parameters on shafting vibration characteristics can be considered. It also can adopt different mathematical methods to calculate and compare with simple mass - spring model. This model can have high precision.

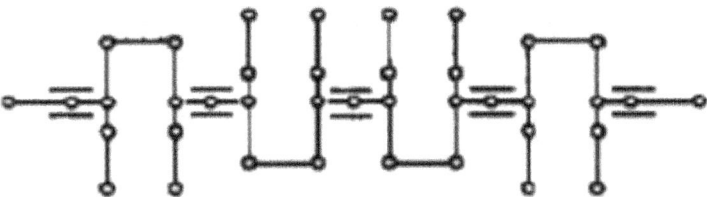

Figure. 2. Framework model schematic diagram.

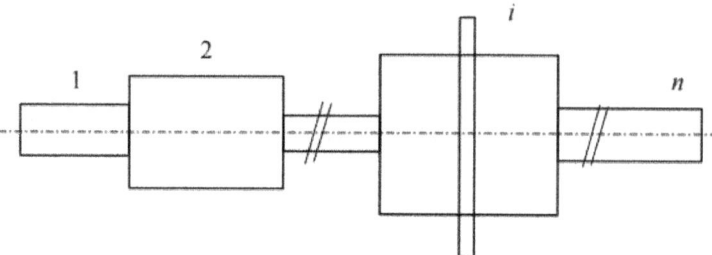

Figure. 3. Multi-diameter model schematic diagram.

Multisegment Concentrated Mass Model

This model is similar to the simple mass model in essence. However, it can be separated into dozens to hundreds sections according to the structure characteristics upon analysis demand. It can calculate high order torsional vibration frequencies that can't be determined by simply mass model, and also avoids the large amount of computation that required in the calculation of continuous mass model. Thus it has been widely used.

Soft Body Dynamics Model

In the calculation of flexible multi-body dynamics, flexible body is described as modal flexible body. A flexible body contains a series of modals. In the breakdown steps, each model unit requires obtaining system state variables and calculating the relative amplitude of each characteristic vector, then using linear superposition principle to integrate node deformation of each time step to reflect total deformation of flexible body.

Other Axis Modeling Methods

In recent years, with the further study of shafting vibration, many new modeling methods came up in the engine industry and other related industries.

Continuous beam model was used in the crankshaft load calculation by Li Renxian [15]. The crank and conrod were equalized to the concentrated force acting on a non-equal continuous beam, and all kinds of force were also equalized to simplify. The author analyzed various loads of crankshaft and its changes in an operation cycle comprehensively. This is a simplified model force shown in figure 4. Of course, in order to calculate simply, the author treats both gas load and centrifugal force function load as concentrated loads. If they were expressed as some forms of distributed loads, the calculation might be more accurate. Calculation model may also adopt continuous beam to make it more close to the actual situation of crankshaft.

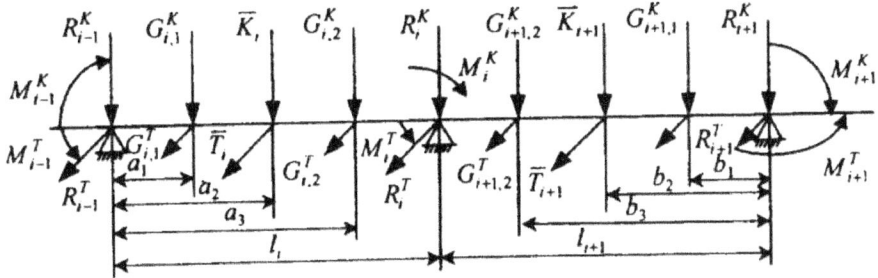

Figure. 4. Continuous beam model schematic diagram

Gu Yujiong[16] used the four-terminal network model in calculating torsional vibration. The author starts with motor control equations and its general solution, equalizing torsional vibration system to the four-terminal network model based on the principle of electromechanical analogy; then adopts mechanical impedance method to obtain frequency equation controlling torsional vibration according to the input impedance of the system and the resonant characteristics, and then obtains the natural frequency, vibration mode and stress distribution, etc. of each order. Four-terminal network model is an accurate low-order model, whose algorithm is convenient and fast. The physical significance is obvious to analyze the mechanical impedance of the system, thus it is a good attempt to model torsional vibration. He Shanghong and Duan Jian [17] also used network method in calculating torsional vibration, and based on dynamic in the process of modeling, which was also a kind of deepening of the method.

Through the analysis of different vibration mathematical models, Xiang Jianhua [18] proposed a graphical modeling method based on system matrix method solving the axis of torsional vibration. The modeling method only requires users providing original torsional vibration mechanics model, and is not restricted by axis branches and modeling scale. In actual implementation, torsional vibration module is divided into two kinds of module unit in this method, which can be used to build various kinds of torsional vibration mechanical model. Model topology relation can be generated through the module traverse and equalisation conversion of the torsional vibration model, and finally the system integration required for solution is integrated.

Axis Modeling Research Direction

Thanks to the development of modern computer, the precise calculation for shaft can be easily realized by finite element method. So for the continuous mass model, the main development direction is how to make the model have better simulation with material object in computer modeling, thus to peer analysis of the model to material object.

As the model parameters (especially rigidity parameters) of a simple mass - spring model are obtained by a large number of experience formula and approximate calculations, as its accuracy is hard to ensure, resulting in rather great error between calculated results and actual machine test results. The reason that theoretical calculation result has lager error is often not because of the calculation method of itself, but lies in the accuracy of the model. Since rotational inertia has a relatively accurate analytic calculation, and the torsional vibration damping of the axis is small, to improve the accuracy of the models, focus should be on the amendment of rigidity parameters. Multi-segment concentrated mass model also has similar problem.

The establishment and derivation of a new model should be based on the amendment with material object, adopting all kinds of similar models used in other industries to derive, thus further perfect engine crankshaft torsional vibration model. The influence of damping should not only be considered, the influence of bending-torsional mixture should also be taken into account. Now, many scholars apply model reduction method [19]used in dealing with torsion vibration of steam turbine and generator as well as the method [20] used in identifying parameters of experimental data to the calculation of the engine torsional vibration. This is also the evitable trend of torsional vibration integration.

SOLVING METHOD OF TORSIONAL VIBRATION OF INTERNAL COMBUSTION ENGINE

Common Method of Torsional Vibration

Based on the above-mentioned several shaft models, the common methods and algorithms solving torsional vibration for multi-freedom free vibration calculation include Holzer method, system matrix method and transfer matrix method, etc. The methods for multifreedom forced vibration calculation include energy method, amplification coefficient method and system matrix method, etc. With the development of computer technology, the traditional manual calculation has been replaced by computers gradually, while some common calculation methods of torsional vibration emerged, such as modal analysis method and finite element method, etc. Various methods are described below.

Holzer Method

Holzer method [1, 2] is always a classic and effective solution in "free-free" system of power machine. The Holzer form method or the Tolle form method derived from its basic principle are often used in engineering. The Holzer method, widely used today, is a numerical calculation method and corresponding calculation program derived from its principle. The advantage of this method is clear physical concept. From its scientific name, this method can be called "method of sum of torsion moments". The basic idea is: the sum of inertia moment of each lumped mass (disc) should be zero when the shaft doing free vibration without damp, that is

$$\sum I_k \ddot{\varphi}_k = 0$$

Due to the characteristics of simple harmonic oscillator, the relation between the displacement α_k and the acceleration $\ddot{\varphi}_k$ of each inertia I_k is:

$$\ddot{\varphi}_k = -p^2 \alpha_k$$, namely,

$$\sum I_k p^2 \alpha_k = 0$$.

This is the foundation of Holzer method.

This method is effective in estimating low order torsional vibration frequency in initial design stage. This method has simple algorithm and is easy to use, thus is widely used in actual engineering. But its higher-order calculation has lower precision and is time consuming.

System Matrix Method

System matrix method [1, 2, 21] is a method using each parameter matrices of torsional vibration equation of the axis to solve characteristic root to calculate torsional vibration. All methods which can calculate the eigenvalue and eigenvector of the matrix can be used to calculate the free torsional vibration of multi-mass system.

The basic principle of this method is: for more freedom vibration equation:

$$\mathbf{I}\ddot{\varphi} + \mathbf{K}\varphi = 0$$

By assuming the form of solution and input them into equation, the following can be obtained:

$KA = \omega_n^2 IA$ let $\lambda = \omega_n^2$ and $H = I^{-1}K$,

now the system matrix of $W_M = W_C$ can be obtained. So free vibration calculation can come down to the question of solving characteristic equations $|D| = 0$.

System matrix method is widely used, not only in free vibration solution, but also in the solution of forced vibration. However, it's generally only suitable for solving low frequencies and its accumulative error would be bigger for calculating high frequencies.

Transfer Matrix Method

The transfer matrix method [16,22] is a commonly used method for analyzing various vibration problems, which was first introduced by Holzer to analyze crankshaft vibration and calculate the inherent frequency of undamped-free vibration of the shafting.

The basic concept of transfer matrix method is: decomposing the studied system into several two-terminal elements with simple mechanical properties, and building relation between the state vectors of the two terminals of one component by transfer matrix. Then, connect all components one by one, and multiply them together to obtain and solve the transfer matrix. Internal combustion engine shafting, according to its composition configuration characteristics, can be divided into three kinds of components: inertial disks - viscous damper components, elastic elements and even elastomer shaft section components.

The advantage of vibration calculation by transfer matrix method is that the order of transfer matrix will be not affected by the increased unit number, namely, the dimension of matrix will not increase with the increase of the freedom degree of the system, and the calculation method of each order vibration mode is identical. So with simple calculation, convenient programming and less memory for calculation and less time consumption, this method is widely used in the analysis and research of crankshaft vibration. However, when analyzing complex shaft with many freedom degrees by this method, due to the error accumulation of the transfer matrix, the calculation accuracy will decrease, thus the precision of higher frequencies computation is relatively low.

Energy Method and Amplification Coefficient Method

Both energy method and amplification coefficient method[2,23] belong to the resonance calculation method of forced torsional vibration, which are basic and the most important calculation methods of torsional vibration before electronic computer popularized. They are still widely used at present. The basic principle of energy method is that the input energy of the exciting moment within a system vibration period is completely consumed by system damping, namely, $W_M = W_C$. Amplification coefficient method was proposed by Tuplin in 1930's for resonance calculation, and then was further developed, becoming a guiding method that the Shipping Standard of British Lloyds Register recommends.

Modal Analysis Method

The basic thought of modal analysis method [24-25] is to decompose complex multi-freedom system into several sub systems. Firstly, when analyzing, compute the several lower modes of each sub system, then assemble each sub system into an integrated motion differential equation set according to displacement compatibilities or force balance relations between adjacent sub systems to derive comprehensive eigenvalue problem of shrinkage of Freedom Degree., thus work out the inherent frequency A vibration mode and response of the system. Since modal analysis method reduces the freedom degree of the system, the time consumption and memory for calculation are significantly reduced compared with finite element method. If the sub systems are divided reasonably, its calculation precision is also satisfactory. In addition, modal analysis method can also combine with experimental research [26] to obtain system vibration modal parameters by measuring the transfer function of shaft vibration, e.g., natural frequency, vibration mode, damping, modal inertia and modal rigidity, etc.

Finite Element Method

Finite element method [1,2,7-10] is a numerical calculation method for solving mathematical physics equation based on variation principle. Its basic thought is to regard complex structure as finite set of discretized units. Each unit is connected into a unity through the common point of the neighboring units, namely, "joints". Take each unit as a continuous component and joint displacement as generalized coordinate. To establish torsional vibration mechanics model of the shafting of internal combustion engine, we need to define which units are selected as well as load positions and sizes, etc. Finite element method is currently accepted as with the highest calculation precision for torsional vibration calculation.

Substructure Analysis Method of the Torsional Vibration of Systems with Branch Shafts [27]

In the torsional vibration analysis of shaft systems with branches, the main commonly used methods are transfer matrix method, matrix iterative method and system matrix method, etc [28-31]. But these methods are mainly used to analyze straight string structure or a particular branch. Their calculation efficiency is relatively low for the whole branching structure system. In general, substructure method has already formed systemic theory [32-33], whose basic idea is to divide large and complex structure system into several substructures and calculate the dynamic characteristics information of each subsystem by finite element method, analytical method and experimental method, and then integrate them into the dynamic characteristics of the whole structure system. But substructure method is used less in torsional vibration analysis of shaft systems with branches. Representative method is dynamic substructure matrix method. This method requires working out the compatible relation among all substructures in substructure integrating, which leads to the complicated and tedious modal synthesis process in case the amount of divided substructures is large. Thus this method has certain limit in solving complex shaft systems with branches.

Chu Hua [34] and Z P Mourelatos [35] combined substructure method with transfer matrix method in torsional vibration calculation, which became a new migration substructure method, and the new method was compared with finite element method.

According to the structure of shaft systems with branches, Li Shen and Zhao Shusen [27] put forward a method that divided substructures and integrate step by step based on gear meshing form to obtain the torsional vibration inherent characteristics of the whole system. This method was also used in analyzing the torsional vibration inherent characteristics of the structure of the main transmission system with branches of 650 rolling mill. The comparison with the results calculated by other methods shows the feasibility of substructure graded division and stepwise integration method. This principle has expanded the application scope of the substructure modal synthesis in solving the torsional vibration of shaft systems with branches and effectively solved the problem of complex system, thus provided a good idea for solving the torsional vibration of shaft systems with branches.

New Research Methods for Torsional Vibration

In recent years, the number of scholars engaged in research of vibration has continuously increased and new algorithms kept on emerging continuously, such as elastic wave propagation method, eigenvector method and frequency analysis method, etc. According to the theory of torsional elastic wave, Bogacz [36] gave out a method to solve torsional vibration dynamic response by torsion wave method. Shu Gequn and Hao Zhiyong [11,37] also presented a new torsional vibration response calculation method based on the theory of torsional elastic wave, whose basic thought is: the torsional vibration of the shafting is caused by the torsional elastic wave propagation along the shaft; elastic wave propagates along the axis forward and back in traveling wave form; when one traveling wave meets with another after reflection or delay, , both waves will stack into standing wave causing torsional vibration if their phases are appropriate. The method can be used to analyze continuous parameter distribution boundary, transient response and steady-state response of the crankshaft axis with transient boundary conditions and other vibration characteristics. Since it only requires solving linear equations in calculation, its computational complexity is small, thus it is an accurate and fast vibration analysis method. According to electromechanical analog principle, Gu Yujiong [16] put forward a four-terminal network method for analyzing torsional vibration. State vector method, proposed by He Chengbing [38] , was widely used in the analysis of torsional vibration. People are also exploring the calculation method of continuous mass model for torsional vibration response of forced vibration. Wang Ke she and Wang Zheng guang[39] used frequency analysis method in the calculation of torsional vibration to combine frequency change with the structural parameters of shafting, which was beneficial to visual analysis. It also worked out analysis mode, resonance frequency and resonant modes, etc.

Research Direction of Shafting Solving Methods

At present, the methods for solving torsional vibration are various and each has its own use. While in general, it shall be developed from the following aspects:

1. Improve the computation efficiency of current methods. For instance, calculation precision is high by finite element method, but its calculation is time consuming and resources occupying, so fewer and dimension-reduced units should be considered in model building when improving this method.

2. Combine the calculation method used in other industries with this direction.

    a. For example, Shen Tumiao [16] mentioned to apply electrical four-terminal network in the calculation of torsional vibration. This is an example of unified calculation method. In addition, integrating all calculation methods to construct a new method is also one of the research directions.

b. Based on the torsional vibration numerical simulation study of the internal combustion engine shaft with the precise time integration method, Lin Sen [40] introduced and deduced the precise time integration based on Duhamel integral, described in detail the calculation characteristics of this method with example and comparison, simulated the torsional vibration of some type of internal combustion engine, compared the results with the calculation results in literature, and analyzed their similarities and differences briefly, which, to a certain extent, solved the conflict between the accuracy and stability of calculation.

c. Along with the development of microcomputer technology, we can use professional software to analyze torsional vibration of internal combustion engine. Tong-Qun Han [41] introduced the functions and characteristics of engine simulation software EXCITE—designer developed by AVL company, analyzed torsional vibration and vibration reduction of the shaft system based on the software targeting at the problem of a car engine flywheel bolt fracture, and put forward correcting measures.

EXPERIMENTAL STUDIES ON ENGINE CRANKSHAFT

Current Torsional Vibration Measurement Methods

Torsional vibration measurement is an important content in the study of crankshaft vibration. Compared with transverse vibration measurement, the extract and analysis of torsional vibration signals are both difficult. There are basically two kinds of torsional vibration measurements: contact measurement and non-contact measurement. The former installs sensor (such as strain gauge, accelerometer, etc) on the shaft, and the measured signal is transmitted to instrument by collector ring or radio signal. Non-contact measurement commonly uses "measuring gear method", which uses shaft encoder, gear, or other repeated structure to measure angular velocity in homogeneity to measure torsional vibration. If designing Doppler test method properly, laser can also be used to measure torsional vibration. The followings are introduction to various methods and analysis of their error sources and applications.

Mechanical Measurement

Geiger torsional vibration analyzer is a typical mechanical torsional vibration measuring instrument [2, 23] and was used in torsional vibration study in the earliest stage. This instrument is designed dexterously, whose signal acquisition and signal record are both realized by mechanical devices, simple and practical. It is widely used in the study of torsional vibration. DVL torsional vibration instrument also belongs to this type of torsional vibration instrument. However, the torsional vibration of this method is transmitted to measuring head shelf by belt, the belt elastic vibration will cause distortion. The response bandwidth of mechanical measuring system is very limited, and also because disc springs cannot be too soft, so very low frequency torsional vibration cannot be measured. In addition, the measured signal cannot be analyzed directly by means of modern analytical instrument, thus it has gradually been eliminated.

Contact Measurement

Contact measurement [42] is to install sensor (such as inductance, strain gauge, etc.) on crankshaft directly. The measured signals are transmitted to analytical instrument by collector ring or in radio frequency manner. To monitor the dynamic response of shaft or shaft parts (e.g., blade, etc.), the arrangement of strain gauge should eliminate the interference of transverse vibration, and can realize the automatic compensation of influence by temperature. Torsional vibration meters belonging to this measurement method include strain-gauge torsional vibration meter, piezoelectric torsional vibration meter and inductance-type torsional vibration meter, etc. Contact measurement, centered by sensing element, is widely used in the vibration test of internal combustion engine, thanks to its high sensitivity, wide frequency response range and convenience for measured signal record and analysis. But this kind of measuring device system itself has certain rotational inertia, which will inevitably impact on the system under test in measurement. In all kinds of contact measurement, measurement devices, such as sensors, are required being installed on the shaft, which sometimes has to destroy the original shaft structure. This is not allowed in many cases.

Non-Contact Measurement

The measurement device of non-contact torsional vibration measurement [43,44] is not installed directly on the crankshaft, but collects torsional vibration signals through photoelectric and magnetoelectricity conversion by code disc, gear or other indexing structure on the crankshaft. These kinds of method are based on the principle of "gear testing". When the shaft is rotating, the teeth structure installed on the shaft can induce bell shaped pulse leveling signal sequences on the sensor, whose amplitude and phase might carry the information on axial torsional vibration, which is demodulated by phase detectors into torsional vibration signals. Torsional vibration meters belonging to this measurement methods include TV - l torsional vibration meter of British

Econocruise Company, VED - 233A torsional vibration meter of American Shaker Company and DTV - 88 torsional vibration meter[45] developed by Shanghai Institute of Electrical Equipment, etc. Non-contact measurement method does not need installing special devices on the shaft, but uses the existing shaft repeated structure, whose measurement preparations is less, and measurement process does not interfere with the normal operation of shaft. It's especially suitable for the long-term monitoring of torsional vibration. At present it has become a major means of torsional vibration measurement.

Laser Measurement

Laser Doppler Torsional Vibration Measurement Technique [46-48] is put forward and developed from fluid velocity measurement. When laser beam irradiates on shaft surface, the linear velocity of shaft surface make scattered light produce Doppler frequency shift. The transient angular velocity of shaft represents the transient value of frequency shift volume of the instantaneous axis. Torsional vibration is obtained by removing dc component. 2523 torsional vibration meter launched out by the Denmark B&K Company was a typical representative of Doppler laser torsional vibration meter. In 1994, Ge Weijing [49] and others from Tianjin University applied laser Doppler velocimetry on the torsional vibration measurement of internal combustion engine shaft. Only a smooth section on the surface of the shaft is required, and measuring point is easy to be set up. This method can realize absolute measurement and measurement datum is not required to be specially established. However, since the transverse vibration of the shaft and the form and position errors of the shaft section directly affect measurement precision, it is rather difficult to improve its accuracy.

The Latest Torsional Vibration Measurement Method

Wang Ting and Cheng Peng [50] introduced a kind of digital measurement system of crankshaft using PC computer. The measurement system consists of angle encoder, self made count plate, PCL724 digital input/output card and PC computer, etc. installed on the tested crankshaft. Angle encoder is crankshaft angle sensor with high precision, and is connected with crankshaft by flange. The grating disc fixed in the angle encoder has two reticules, e.g., outer and inner rings. The outer ring is a uniform reticule, which can produce CDM signals, and the inner ring is a TRIG reticule for judging tdc signal. The light emitting components in the angle encoder are two infrared light emitting diodes, and there are two infrared light receptors respectively corresponded with CDM reticule and TRIG reticule. When angle encoder operates together with crankshaft, a TRIG signal will be outputted in each rotation, and a series of CDM square-wave pulse signal will be outputted. Thus, the crankshaft torsional vibration can be directly reflected on the time width of the CDM square-wave pulses outputted by angle encoder. Count each CDM pulse width with frequency division by high count circuit board. The counted data is inputted into PC by parallel data I/O card PCL - 724. Then crankshaft torsional angle can be obtained after program processing. The measurement system measures torsional angle directly, thus it's with convenient measurement, high precision and simple process.

Research Direction of Torsional Vibration Measurement of Internal Combustion Engine

The focus and future development of torsional vibration measurement is to improve its accuracy and real-time performance to realize the torsional vibration monitoring of internal combustion engine in operation, especially the monitoring of severe torsional vibration caused by emergencies, such as the severe torsional vibration by transient large torque incentive resulted from cylinder flameout. At the same time, eliminating the interference of lateral vibration and establishing reliable measurement datum are still the problem requiring to be solved. Finally, the problem of system calibration should also be solved.

THE LATEST RESEARCH DIRECTION OF TORSIONAL VIBRATION OF INTERNAL COMBUSTION ENGINE

The traditional research methods of torsional vibration cannot meet the needs on the precise study. In recent years, many scholars have continuously broadened research field and scope to further explore the various problems of shaft torsional vibration, making the research on torsional vibration closing to ideal level unceasingly. Some main research directions of torsional vibration in recent years are introduced as follows.

Nonlinear Research

With the further research on torsional vibration of the shaft, many nonlinear vibration problems are met [51-55]. At the same time, crank shaft is a complex nonlinear system, thus it often needs to consider all sorts of complex nonlinear factors to construct a model that can reflect actual system. However, current relevant studies are mostly on single degree-of-freedom nonlinear vibration problems that considers single factor, which obviously cannot meet the need on the accurate calculation of crankshaft vibration. Therefore, it is necessary to further consider the crankshaft nonlinear vibration problems with multiple nonlinear factors, multi-degree-of-freedom and even continuous mass distribution. In the current calculation models, equivalent moment of inertia (constant) is usually adopted to consider the inertia of piston and connecting rod. But in fact, the inertia of engine crank module is:

$$I = I_0 \left[1 - \varepsilon \cos(2\varphi) \right]$$

Where ε is variable inertia coefficient, a value below 1. We can see from the formula, the moment of inertia of crank component is a variable related with rotation angle. Sheng Gang [56] researched on the solution methods of some simplified models of single cylinder engine and established the equation of motion of crankshaft vibration of multi-cylinder engine under the condition of considering variable inertia. At the same time, in literature [1], the problems of variables caused by machining error and assembling error were also considered. In literature [57], forward and inverse Fourier transformation was applied to numerically solve nonlinear torsional vibration system. While Lin Ruilin [58] took the diesel engine shaft with a third-order rigidity component as research object, deduced the calculation formula and numerical calculation formula iterative procedure for solving periodic response of nonlinear torsional vibration by incremental harmonic balance method (IHB). This method is used to solve linear and nonlinear torsional vibration response of diesel engine shaft. Compared with the existing methods, it is more effective to solve strong nonlinear vibration response. What's more, it has virtues of less operation time and accurate calculated result.

The discussion about nonlinear components is mainly concentrated on the non-linear shock absorber, coupling and other components. Literature[59] of as early as 1987 analyzed the nonlinear problems of diesel engine shaft with piecewise linear components (cylindrical spring-loaded buffers), and calculated vibration response of shaft by step-by-step integration method. Gong Xiansheng[55] introduced theoretical and experimental research on the calculation method of steady state vibration response of marine propulsion shafting with hysteretic nonlinear coupling subjected to eccentric mass exciting force action. Farshidianfar [60] solved nonlinear problems of driving shaft by substructure modeling, and compared the results with the results of whole structure modeling. The research on nonlinear torsional vibration has made many important achievements [50] . But so far, the majority of nonlinear torsional vibration problems are still analyzed by some approximate methods or by ignoring nonlinear factors, in most cases, the results obtained have greater errors compared with actual results. Therefore, there are still many problems waiting to be solved in further exploring the nonlinear problem, mainly including:

- The modeling, system parameters identification method and test of complex nonlinear torsion vibration problems;
- Accurate solving methods for multi-degree-of-freedom strong nonlinear torsional vibration problems;
- Self-excited vibration of complex nonlinear torsional vibration system;
- Decoupling, numerical calculation and optimization methods of complex nonlinear structure.

Coupling Vibration Analysis

Torsional vibration of shaft has huge harm on the system, so people paid attention to and researched on it at very early period. However, many phenomenon produced by vibration in practice need to take longitudinal/bending/tortional vibration together into account. The bending vibration caused by unbalanced mass has certain weight on the torsional direction and can couple to the torsional vibration; on the other hand, torsion also has certain weight on horizontal and vertical directions, thus couples to the bending vibration. In recent years, significant progress has been made in the aspects of theoretical calculation method and testing technology of longitudinal/bending/tortional coupling vibration.

Li Bozhong [62] discussed about the axial vibration problems caused by torsional vibration and established a relatively simple analysis model for this kind of model. In the following two literatures[63,64] of same series, the longitudinal twist coupling vibration was tested and further analyzed and the coupled vibration model was established, and the model calculation was compared and analyzed with actual measurement. In paper [65], the author put forward a kind of spring - mass model with non-linear rigidness being used in calculating torsional - vertical coupled vibration of engine shaft. It explained the doubled frequency problem of the torsional - longitudinal coupling, and also revealed the presence of quadruple frequency and octuple frequency in the longitudinal - torsional coupled response. It is more reasonable than just simply giving an assumption doubled-frequency excitation torque in the right of the motion equation. Zhang Yong and Jiang Zikang[66,67,68] adopted distributed mass model in analyzing bending - torsional coupling vibration of shaft, which divided the actual unit shaft system structure into several sections with equal diameter according to orders in simplifing, treated each segment as continuous mass, and listed the vibration differential equation of each segment, then united them to solve. Finally, some results of the analysis for bending - torsional coupling vibration of shaft by numerical method were given. In literature [69], system matrix model was established for longitudinal twisting coupling vibration of shaft, whose general rule of coupling vibration was studied based on the calculation and analysis of the practical examples of longitudinal twisting coupling vibration of shaft. In this paper, the test equipment used for measuring coupling vibration is only eddy current sensor for non-contact measuring the condition of axis vertical vibration. In contrast, multi-dimensional measurement is relatively rare.

Okamura[70] and Shen Hongbin[71] used the longitudinal / bending / tortional vibration test device in all research processes of shaft vibration. This kind of measuring device can acquire three-dimensional vibration signals simultaneously. As shown in figure 5, an electromagnetic sensor (measuring torsional vibration signal) and three acceleration sensors (of which, one measuring longitudinal vibration signals and the other two measuring bending vibration signals) are mounted on its shell. It shows that testing technology has also been developed from single parameter measurement to multi-parameter measurement method.

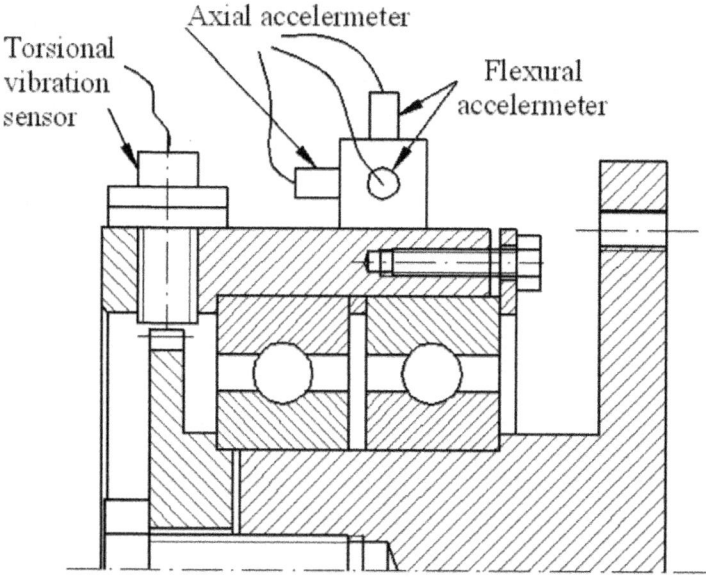

Figure. 5. Three-dimensional torsional vibration measurement schematic diagram.

For the research of torsional vibration with coupling vibration and transversal vibration, the current research level is far insufficient. Especially in the study of theoretical models, traditional method cannot unify the physical model and mathematical model of coupling vibration simultaneously. In view of this, the future research on coupling vibration can roughly focus on the following aspects:

- Mechanism research on coupling vibration and relay relation of mutual excitation;
- Model unification and solve using universal algorithm;
- Precise treatment of measuring equipment, long distance measurement and the implementation of long-term test, etc.

The Analysis of Torsional Vibration Response Based on Multi-Body Dynamics of Soft Body Crank Shaft

The forces on engine are very complex. Traditional analysis method is to calculate rotation inertia and reciprocating inertia produced by each force based on the motion analysis of each component, then combine them with the maximum combustion pressure of gas to solve the force on the main body and excitation force of shaft vibration. This is a very complicated process[73]. By using mechanical system simulation software ADAMS, by establishing crankshaft multi-body dynamics model including pistons, connecting rod, crankshaft and flywheel, we can not only calculate the motion law and the force among each component, but can also further analyze balance and vibration. Due to the interaction between inertial load and transverse bending deformation of shaft and the coupling behavior with lubrication problem, the bearing load problem based on rigid body dynamics becomes complicated, and there exist errors in calculation precision. If transforming engine crankshaft into flexible body, the tiny deformation can guarantee the completely accurate dynamic equation to deformation generalized coordinates first-order items. In order to sufficiently study the effect of crankshaft flexible body on the calculation results of dynamics, based on the finite element analysis of crankshaft system and by establishing rigid-flexible coupling multi-body dynamics system model with multiple degrees, Liang Xingyu and Shu Gequn[72] analyzed the torsional vibration response of crankshaft system that constitutes main flexible body, and obtained the time history response of system dynamics, and then made assessment on the power quality and safety of the system. Then it measured the torsional vibration of the crankshaft free end of an inline four cylinder diesel engine with a newly developed test device. Through calculation and comparison between sub-harmonic analysis of test results, both reflected higher equality, and explained the correctness of rigid-flexible coupling multi-body dynamics system model.

The Method of Compensating Divisional Error in Shafts Torsional Vibration Measurement and Program Implementation [74]

Now non-contact measurement method are generally used for torsional vibration measurement, namely, by using repeated structure in the shaft, pulses are produced in noncontact sensor, and the interval dimension reflects the transient angular velocity dimension of the shaft. The shaft torsional vibration information can be obtained by processing interval data of the pulses. When measuring torsional vibration by this method, the indexing error of the shaft repeated structure directly influences the precision of measurement results. If indexing error is very great, the measurement results will have serious distortion. In the measurement of torsional vibration by noncontact measurement method in practice, selected sensors mainly include photoelectric encoder, hall sensor and photoelectric sensor. The three kinds of sensors in practical measurement have their advantages and disadvantages. Photoelectric encoder has large indexing number and small indexing error, so its measurement is more accurate. But its installation is inconvenient and it requires using shaft coupling to connect with measured shaft. If encoder shaft is eccentric with measured shaft after installation, transmission eccentric error will be introduced[75]. Hall sensor requires gearing disc with equal division to measure. Since the teeth number of equal division disc is usually less than the indexing number of encoder, and gear disc has certain indexing error in processing, its accuracy is lower than that of encoder. But its installation is convenient. It can be directly installed on the measured shaft, or can directly measure by gear disc on the shaft without additional modification on the shaft. The use of photoelectric sensor is of the most convenience. It needs only uniformly pasting a certain number of reflective strips on the component with circular surface of the shaft. If the shaft is very thick, those reflective strips can be directly pasted on the shaft. However, currently, the reflective strips can only be manually pasted. Great degree error will definitely occur leading to the distortion of measurement results. So in the actual torsional vibration test, if the indexing error of selected repeated structure cannot be ignored, such as selected manually pasted turntable of reflective strips, test results should be dealt with to compensate for the effects of indexing error, then correct torsional vibration information can be calculated. Guo Wei-dong [74] described the compensation principle of indexing error in detail and listed the compensation program of indexing error compiled based on LabVIEW. So we could find out from test results that the data curves after the compensation of indexing error became smooth, and the effect that indexing error on measured results is obviously reduced and test result is more accurate.

CONTROL TECHNOLOGIES IN TORSIONAL VIBRATION OF INTERNAL COMBUSTION ENGINE

For the internal combustion engine with reciprocating motion, due to the property of periodical work, the torque on the shaft is a periodic compound harmonic torque, and then forms excitation source. When the frequency of the excitation source is equal to the inherent vibration frequency, resonance phenomenon will occur, and torsional vibration will be subjected to huge dynamic amplification effect, then the torsional stress on the shaft greatly increases, leading to various accidents on the shaft, and even fracture. These are the causes and consequences of torsional vibration. To avoid the destructive accident of torsional vibration of internal combustion engine, it's not only required to conduct detailed calculation of torsional vibration in design phase, torsional vibration measurement is also required timely after manufacturing completion. This can not only check and modify the theoretical calculation results, but also detect and so as to solve the torsional vibration problems promptly. Based on the above analysis, main vibration control technology includes two parts: study on the avoidance of vibration and on shock absorber.

Study on the Avoidance of Vibration

If great torsional vibration does exist on internal combustion engine according to the calculation of and actual test on torsion vibration, proper measures shall be taken to avoid or remove it.

There are a lot of preventive measures for avoiding torsional vibration [61], classified roughly into the following two methods.

Frequency Adjustment Method [2, 76]

According to torsional vibration characteristics, when the frequency of excitation torsional vibration is equal to some inherent frequency n_w of torsional vibration system, extremely severe dynamic amplification phenomenon will occur, namely resonance phenomenon, thus the possibility of $w = n_w$ shall be avoided, i.e., avoidance of the most severe conditions of dynamic amplification means the possibility of avoidance of all consequences caused by excessive torsional vibration. The basic concept of this method is that let w actively avoid n_w. The main measures of this kind of method include: inertia adjustment method and flexibility adjustment method, etc. By adjustment, let the natural vibration frequency of the system itself avoid excitation frequency. Reduce vibration stress to be within the instantaneous allowable stress range, thus avoid the damage on engine by bigger torsional vibration. This method is one of the most widely applied measures in torsional vibration prevention measures, not only because of it being a simple and feasible measure, but also because of it being effective and reliable when meeting the requirement of frequency modulation. But its disadvantage is small scale of frequency modulation, which restricts its practical application.

Vibration Energy Deducing Method [23]

Incentive torque is the power source causing torsional vibration. Since the input system energy of incentive torque is the source of maintaining torsional vibration, if the vibration energy of input system can be reduced, the magnitude order of torsional vibration can also be reduced immediately. One way is to change the firing sequence of internal combustion engine. When the dangerous torsional vibration is deputy critical rotation speed within machine speed range, this method might be used to reduce the dangerous torsional vibration and reduce the risk degree. The second method is to change crank arrangement. Deliberately choosing unequal interval firing in multi-cylinder engine and appropriately choosing crank angle to change crank arrangement can let some simple harmonic torsional vibration in any main-subsidiary critical speed counteract mutually to avoid dangerous torsional vibration. The third method is to choose the best relative position between crank and power output device, make the disturbance torque between them counteract mutually, which can reduce the torsional vibration of the crankshaft.

Impedance Coordination Method

Considering the complexity of solving above problems by the conventional dynamics method, energy wave theory can be used to solve this problem. According to energy wave theory and by coordinating the impedance of various component loops, resonance can be avoided to realize the target of reducing vibration intensity. Impedance coordination method can modify the inferior design in design phase, or design directly correct transmission shaft system, to ensure the shaft working with sound dynamic characteristics without resonance and reducing dynamic load.

Study on Shock Absorber

As is known to all, engine installed on shock absorber can greatly reduce the vibration transmitted to the foundation. Likewise, torsional vibration can also be eliminated before it reaches the foundation. If vibration reducing device is installed on the front head of the crankshaft of the engine, then shock absorber will absorb the torsional vibration of rotating shaft generated by engine. It shows the important role of shock absorber in internal combustion engine system. The technical requirements on shock absorber are very high, mainly including: elastic material strength should be reliable in use and storage, the fixation with metal should be firm, rigid fluctuation range in installation stage should be small, and technical characteristics do not change with time.

Now, main shock absorbers include the following kinds: dynamic shock absorber, damping shock absorber and dynamic-damping shock absorber.

Dynamic Shock Absorber [2, 23, 77]

This kind of shock absorber is connected with crankshaft by spring or short shaft. By the dynamic effect of shock absorber at resonance, an inertia moment with the size and frequency same with excitation torque, but direction opposite to excitation torque is produced at the vibration reduction location to achieve the purpose of vibration reduction. This kind of shock absorber doesn't consume the energy of the shaft. They can be divided into two types: one type is constant fm dynamic shock absorber, namely undamped elastic shock absorber, shown in the schematic of figure 6, and the other type is variable fm dynamic shock absorber, such as undamped tilting shock absorber, drawing as shown in figure 7.

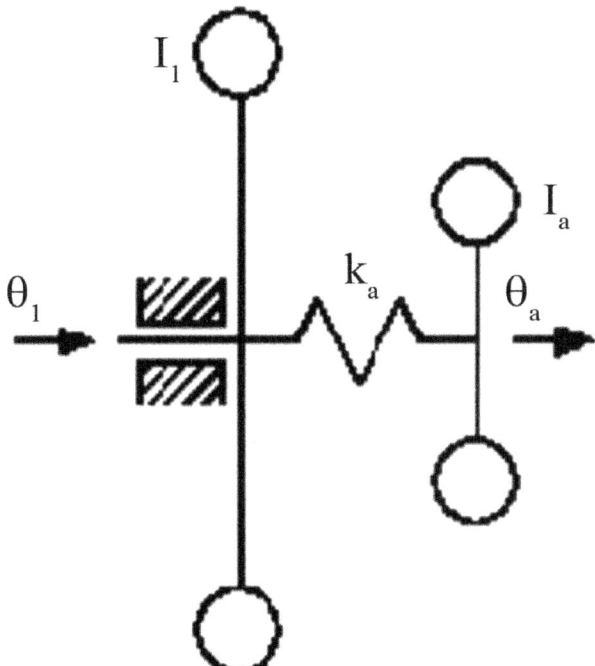

Figure. 6. Undamped elastic shock absorber schematic diagram.

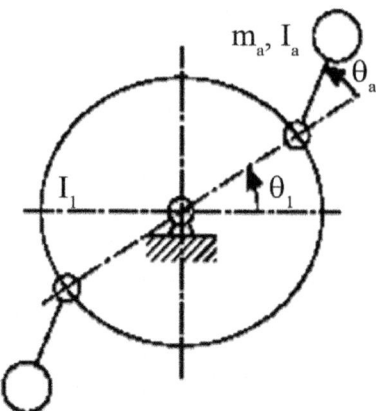

Figure. 7. Undamped tilting shock absorber schematic diagram.

Damping Shock Absorber

Damping shock absorber achieves the purpose of vibration reduction by damping consuming excitation energy (shown in schematic diagram 8). The main type is silicone oil damper[78,79], whose shell is fixed to the crankshaft, high viscosity silicone oil is filled between ring and shell. When the shaft is under torsional vibration, the shell and the crankshaft vibrate together, and the ring moves relatively with the shell due to inertia effect. Silicone oil absorbs vibration energy by friction damper, thus reduce vibration for the torsional vibration system. This kind of shock absorber is widely used with simple structure, good effect of vibration isolation and reliable and durable performance.

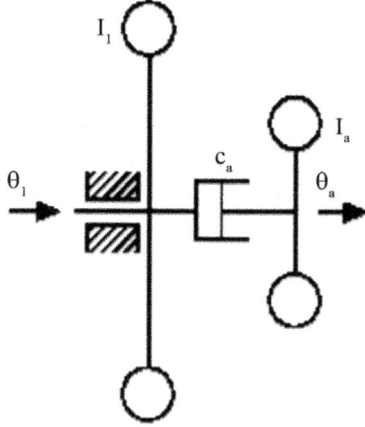

Figure. 8. Damping shock absorber schematic diagram.

Dynamic-Damping Shock Absorber

Dynamic damping shock absorber features both of the above effects, such as rubber flexible shock absorber [80], rubber silicone oil shock absorber, silicone oil spring shock absorber [81], etc., shown in schematic diagram 9. Theoretically, the effect of dynamic damping shock absorber is the best, since it can not only produce dynamic effect by elastic, but also consume excitation energy by damping. But the elastic elements, such as springs and rubber, etc., that connect the shock absorber and crankshaft often work under great amplitude and high stress, thus the process is relatively complex and the cost is higher.

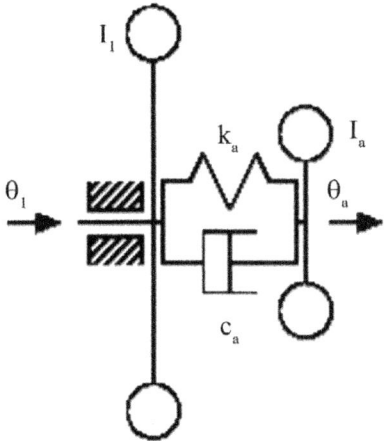

Figure. 9. Dynamic-damping shock absorber schematic diagram.

Study on New Shock Absorber

Along with the deepening of the research on shock absorber, many new shock absorbers have appeared. Several typical kinds are listed as below:

Yan Jiabin [82] proposed an elastic metal shock absorber, which fixed two disks connected with elastic materials at the ends of the crankshaft. disks are tightened in the way that one disk rotates in the opposite direction of the other disk (see figure 10). If loosen both disks simultaneously, they will complete torsional vibration with low amplitude, till stop. At this moment, one section of the shaft rotates in one direction, and another section of the shaft rotates in the other direction. In this case, one end face of the crankshaft will produce displacement. Due to the effect of vibration absorption, the vibration will proceed with reduced amplitude but constant speed, which depends on the internal friction or delayed quantity of elastic material. There are three kinds of elastic shock structure of shock absorber: welding metal elastic elements, combination elastic metal components and welding-combination elastic metal elements. Welding-combination elastic metal elements consist of driving and inertia members that connect each other with elastic material. This kind of shock absorber is suitable for application with simple structure and convenient maintenance.

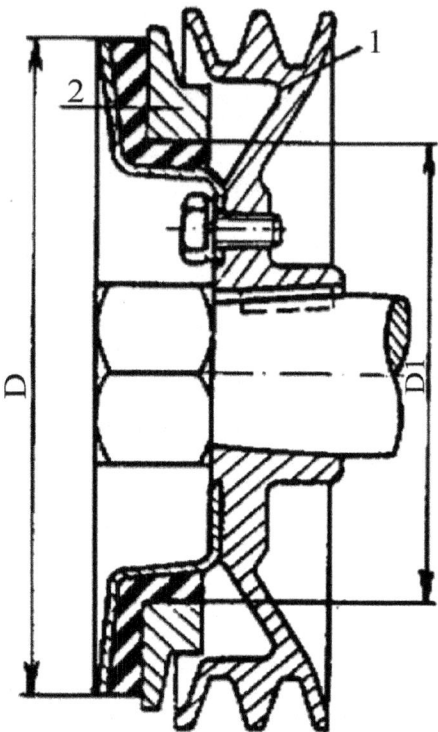

Figure. 10. Monolayer thin-type elastic metal shock absorber.

Huo Quanzhong [83] and Hao Zhiyong[84] introduced the research on driving control shock absorber. Figure 2 is the diagram of the shock absorber. The shock absorber itself is similar to a dc motor, whose stator and the shell of the shock absorber compose as a whole entity, and rotor is connected with the shell by radial leaf spring, forming a dynamic shock absorber. The shell of the shock absorber is fixed on the main vibration body. According to the conditions of main vibration body, the regulation apparatus produces control signals with fixed size, phase and frequency, which, by power amplifying, make armature generate control torque (namely, electromagnetic torque). Active torsional vibration shock absorber is feasible both in theory and practice. What's more, its damping effect is better than that of dynamic shock absorber.

Liu Shengtian [85] proposed a double-mass flywheel torsional vibration shock absorber, which was a new type of torsional vibration shock absorber occurred in the middle of 1980's. The double-mass flywheel torsional vibration shock absorber at early period was to remove torsional vibration absorber from the clutch driven plate, place it among engine flywheels, thus double-mass flywheel torsional vibration shock absorber was formed. The basic structure of double-mass flywheel torsional vibration shock absorber has three major parts, i.e. the first mass, the second mass and the shock absorber between the two masses. Relative rotation can exist between the first and the second masses, which are connected with each other by shock absorber.

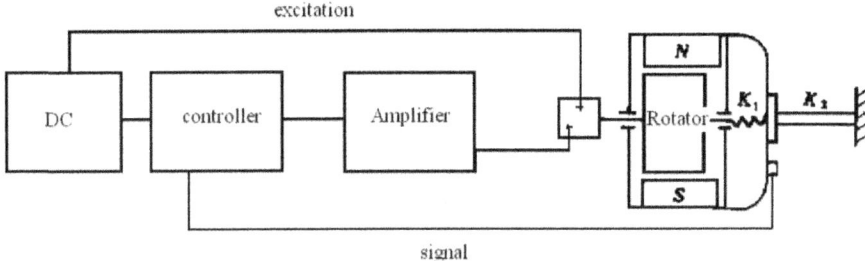

Figure. 11. Active control shock absorber functional diagram.

Double mass flywheel torsional vibration shock absorber can very effectively control torsional vibration and noise of automobile power-transmission system. Compared with the traditional clutch disc torsional vibration shock absorber, its effect of damping and isolation of vibration is not only better within the common engine speed range, it can also realize effective control over idle noise. After developing the double-mass flywheel torsional vibration shock absorber, the author also introduced hydraulic pressure into shock absorber and developed hydraulic double-mass flywheel shock absorber [86], which is the latest structure style in the family of double-mass flywheel torsional vibration shock absorber. It lets the technology of car powertrain and noise control of torsional vibration step further into the direction of excellent performance and simple structure.

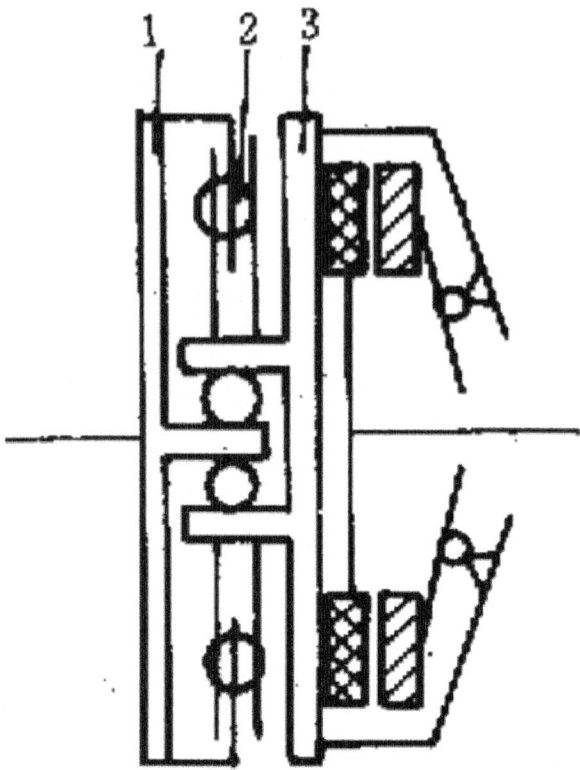

Figure. 12. Double-mass flywheel torsional vibration shock absorber.

M Hosek, H Elmal [87] introduced the design process of a kind of FM tilting shock absorber, which was developed based on centrifugal tilting shock absorber, and was named as centrifugal delay type resonator by the author. Based on the study of centrifugal tilting shock absorber, the author installed a sliding globule between the end of pendulum and the rotary table, thus, when the rotation of the shaft fluctuates, the pendulum will delay duet to the effect of damper, while the sliding globule will coordinate actively with the changes. By this shock absorber, minor disturbance can be quickly completely eliminated; and broadband disturbance, especially the disturbance that obviously increases speed, also can be completely eliminated.

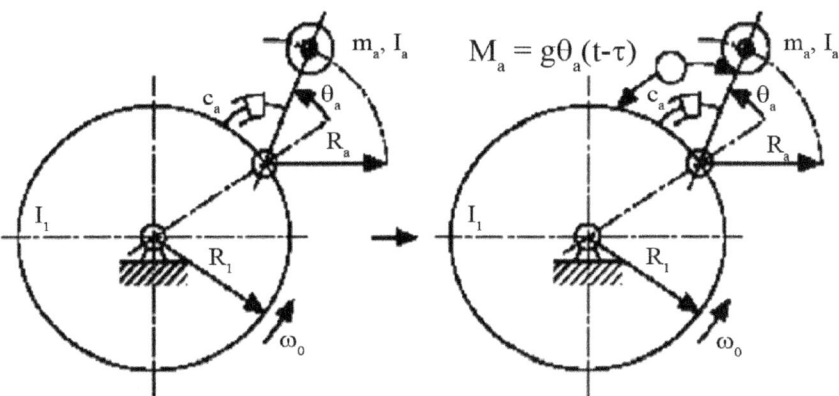

Figure. 13. Centrifugal delay resonator.

Shu Gequn [88,89] presented a research approach of coupling shock absorber. Since torsion vibration is the most dangerous vibration mode in shaft vibration, torsional vibration shock absorber is the main damping device, and for coupling damping, bending shock absorber or lateral shock absorber will be installed on the basis of torsional vibration shock absorber. Through the experimental research, the author concluded that, compared with single torsion vibration damper, after installing bending vibration shock absorber, due to the effective damping act on shaft bending vibration, twist/bending shock absorber can control engine vibration and noise effectively. Normally, the parameters setting of bending vibration shock absorber depends on the bending vibration model of crankshaft, but due to its effect on torsion vibration reduction, the design of coupling shock absorber should consider its damping effect on torsional vibration and bending vibration of the shaft. Shu Gequn [90] investigated the effect of bending shock absorber on the performance of twist/bending shock absorber by theoretical analysis.

Through the above analysis, we can see that torsional vibration absorber is being developed towards the aspects of broadband, high efficiency, being timely, multi-function, etc. So research on torsional vibration shock absorber still has considerable prospect, worthy more efforts from scholars. The following aspects can be studied and explored.

- Study on active control of torsional vibration of internal combustion engine;
- Study on shock absorber with coupling between torsional vibration with longitudinal vibration and transverse vibration, etc;
- Study on integrating torsion vibration absorber with clutch or other components of internal combustion engine;

- Finite element optimization design of torsional vibration shock absorber [91].

UTILIZATION OF TORSIONAL VIBRATION OF INTERNAL COMBUSTION ENGINE CRANKSHAFT

Restricted by various factors, we can only decrease the degree of torsional vibration, while the occurrence of torsion vibration is inevitable. Torsional vibration is directly related to the various incentive factors of internal combustion engine, such as the combustion sequence of various cylinders, the change of crankshaft inertia and the sudden change of loads, etc. So how to use torsional vibration signals to monitor the changes of these quantities is the main purpose of utilizing torsional vibration. The utilization of torsional vibration is mainly embodied in identifying faults by torsional vibration signals[92, 93].

The Progress of Torsional Vibration Utilization

The diagnosis of diesel engine faults by the change of torsional vibration parameters of the shaft is a new fault diagnosis technology. Torsional vibration signals of diesel engine shaft often have strong repeatability and regularity, and fault diagnosis by torsional vibration signals is used to diagnose cylinder flameout. Diagnosing cylinder flameout fault by torsional vibration signal of diesel engine has been developed in recent years. The work process fault of diesel engine cylinder directly affect the changes of torsional vibration characteristics, and such changes of torsional vibration parameters also reflect directly the work state of cylinders. Torsional vibration signals of diesel engine shaft can be used as the basis for fault diagnosis. Ying Qiguang explored this issue at the beginning of 1990's of the 20th century, who thought that diagnosis of the technical condition and fault of diesel engine by the response characteristics of the frequency, amplitude (and phase) and damping of shaft torsional vibration is a new and promising fault diagnosis technology. The author judged cylinder flameout fault by the comparison of amplitude size between normal torsional vibration and in the circumstance of cylinder flameout. This method is convenient and intuitive. However, it requires normally torsional vibration amplitude figure for comparison under the same conditions, so its application is limited [94]. In the application of fault diagnosis by torsional vibration, Lin Dayuan and Shu Gequn studied on the sensitivity of various torsional vibration modals and frequency response characteristics on crack by torsional vibration modal experiment, who recommended modal damping, damping attenuation factor, frequency response function modal and self-spectral modal as the optimum evaluation factors for the crack fault, and further discussed the change law between crack and the above evaluation factors, thus provided an effective method of intermittent diagnosis for the engine stops[95,96].

The Development Direction of Torsional Vibration Utilization

Fault diagnosis by torsional vibration signal of internal combustion engine crankshaft is a new type of fault diagnosis theory. Developing this theory towards new application field is an inevitable trend in internal combustion engine industry. Thus, the development direction of torsional vibration is mainly oriented to broader fault diagnosis fields and continuously make this achievement become more mature and its application become more skilled.

Sub-Conclusions

- Damping technology becomes more mature.
- Multi-function shock absorbers are innovated constantly.
- Application fields of torsional vibration become more and more wide.
- Modern design theory is used unceasingly in control field.

CONCLUSIONS

Research on torsional vibration of internal combustion engine will become more and more deepen with the development of science and technology. Corresponding new research methods will appear in modal building, solving, test and control of the shaft model, making research contents more wide, method more scientific, object more specific and application more direct.

ACKNOWLEDGEMENTS

Authors wish to express their sincere appreciation to the financial support of the NSFC(50906060) and State Key Laboratory of Engines.

REFERENCES

1. Okamura H et. Simple Modeling and Analysis for Crankshaft Three-Dimensional Vibration[J]. Journal of Vibration and Acoustics, 1995,117(1): 70-86.
2. A Boysal and H Rahnejat. Torsional Vibration Analysis of a Multi-body Single Cylinder Internal Combustion Engine Model[J]. Appl. Math. Modeling,1997,21(8):481-493.
3. B O AL-Bedoor. Modeling the Coupled Torsional and Lateral Vibrations of Unbalanced Rotors[J]. Computer Methods in Applied Mechanics and

Engineering, 2001,190: 5999-6008.
4. Bogacz, Szolc T, Irretier H. An Application of Torsional Wave Analysis to Turbo generator Rotor Shaft Response[J]. Trans. of ASME, Journal of Vibration and Acoustics, 1992, 114(2):149-153.
5. Bozhong, Song Xigeng, Song Tianxiang. On Coupled Vibrations of Reciprocating Engine Shaft Systems(Part 2)- Progressive Torsional-Axial Continued Vibrations [J].Transactions of Csice, 1990,8(4):317-322.
6. C T Lee, S W Shaw. The Non-linear Dynamic Response of Paired Centrifugal Pendulum Vibration Absorbers [J]. Journal of Sound and Vibration, 1997,203(5): 731-743.
7. Chen Xi'en. Study on Nonlinear High-damping Dampers in calculation and measurement of torsional vibration[J], China Ship Survey, 1996, 1:26-32.
8. Chen Yong, Lu Songyuan, Gu Fang, Wang Yumin. The Calculation and Analysis of Torsional Vibration Characteristic of Turbine Generator Rotor System. [J]. Turbine Technology, 1999,41(5):276-279.
9. Chu Hua, She Yinghe. A Substructure Transfer Matrix Method in Torsional Vibration Analysis of Rotor System [J]. Journal of Southeast University, 1989,l9(3):87-95.
10. Du Hongbing, Chen Zhiyan, Jing Bo. The Mathematical Model for the Coupled TorsionalAxial Vibration of Internal Combustion Engine Shaft System[J]. Chinese Combustion Engine Engineering 1992,13(2):66-74.
11. E Brusa, C Delprete, G Genta. Torsional Vibration of Crankshafts Effects of Non Constant Moments of Inertia[J]. Journal of Sound and Vibration, 1997,205(2):135-150.
12. Farshidianfar A, Ebrahimi M, Bartlett H. Hybrid Modeling and Simulation of the Torsional Vibration of Vehicle Driveline Systems[J]. Journal of Automobile Engineering, 2001.215(4):217-229.
13. G D Jenning, R G Harley. Modal Parameters and Turbo-Generator Torsional Behavior[C]. IEEE 3rd AFRICON Conference,1992:517-520.
14. Ge Weijing, Wang Weisheng. Small Internal Combustion Engine, 1994,23(2):53-57.
15. Gong Xiansheng Xie Zhijiang Tang Yike. Research on Steady Vibration Response of Shafting with Nonlinear Coupling [J]. Chinese Journal of Mechanical Engineering, 2001,37(6):19-23.
16. Gu Yujiong, Yang Kun, He Chengbing, ect. Simulations on Torsional Vibration Characteristics of Turbogenerators Based on Four-Terminal

Network Method [J]. China Mechanical Engineering, 1999,10(5):540-542.
17. Guo Li, Li Bo. , Assessment of Torsional Vibration Measurement Method[J], Grinder and Grinding. 2000,3:53-56.
18. Guo Wei-dong; Gu Wen-gang; ZhanG Xiao-ling. The Method of Compensating Division Error in Shafts Torsional Vibration Measurement and Implementation with Program [J]. Machine Design & Research, 2007, 23(2): 88-89.
19. Hao Zhiyong, Gao Wenzhi. A Study on Active Control and Simulation Test for Torsional Vibration of Turbogenrator Shaft System [J]. Power Engineering, 1999,19(5): 338- 341.
20. Hao Zhiyong, Shu Gequn. An Investigation on the Torsional Elastic Wave Theory for the Calculation of Crankshaft [J]. Torsional Vibration Response in I. C. Engines. Transactions of Csice, 2000,18 (1) : 29-32.
21. He Chengbing, Gu Yujiong, Yang Kun. Summarization on torsional vibration model and algorithm of turbogenerator unit [J]. Journal of North China Electric Power University, 2003,30(2):56-60.
22. He Chengbing, Gu Yujiong, Yang Kun. The Analysis of the Shaft Torsional Vibrations Response of Turbine-generator Units Based on State-Space Method [C]. Asia-Pacific Vibration Conference, China, 2001.
23. Wang Ke she, Wang Zheng guang. Frequency analysis method of shafting crankle vibration [J]. Journal of Beijing Institute of Machinery, 2001,16(4):11-18.
24. He Shanghong, Duan Jian. Network method for dynamic modeling of complex shafting torsional vibration system [J]. The Chinese Journal of Nonferrous Metals, 2002,12(2): 388-392.
25. Honda Ym, Wakabayashi K, Matsuki K. Torsional Vibration Characteristics of a Diesel Engine Crankshaft with a Torsional Rubber Damper by Using Transition Matrix Method [J]. JSAE Review, 1996,17(1):89.
26. Hua Jianwen, Liu Liren. Study of Measurement by Torsional Vibration Transducer Type MM0071 [J]. Laser & Infrared, 1996,26(3):193-195.
27. Huang Diangui. Experiment on the Characteristics of Torsional Vibration of Rotor-to-stator Rub in Turbo Machinery[J]. Tribology International,2000,33:75-79.
28. Huo Quanzhong, Liang Jie, Hao Zhiyong. Operation Principle and Experimental Study of an Active Torsional Vibration Absorber [J]. Transactions of Csice, 1992,10(4): 329- 334.
29. J Q Pan, J Pan, R S Ming, T Lin. Three-Dimensional Response and

Intensity of Torsional Vibration in a Stepped Shaft [J]. Journal of Sound and Vibration, 2000, 236(1):115-128.

30. Ji Chen, Zeng Fanming. Study on the general model and system matrix method for torsion vibration calculation of complex shafting [J]. Ship & Ocean Engineering, 2006(4): 55-57.

31. K Koser, F Pasin. Continuous Modeling of the Torsional Vibrations of the Drive Shaft of Mechanisms[J]. Journal of Sound and Vibration, 1995, 188(1):17-24.

32. K Koser, F Pasin. Torsional Vibration of the Drive Shafts of Mechanisms[J]. Journal of Sound and Vibration, 1997, 199(4):559-565.

33. Li BoZhong, Chen ZhiYan, Ying QiGuang. The torsional vibration of internal combustion engine crankshaft system[M]. Beijing: National Defense Industry Press, 1984.

34. Li Huizhen, Zhang Deping, Crankshaft Torsional Vibration Calculation by Finite Element Method [J]. Transactions of Csice, 1991, 9 (2) : 157-162.

35. Li Lin, Sheng Jun. Mixed Interface Direct Component Modal Synthesis Method [J]. Chinese Journal of Applied Mechanics, 2005 22 (2): 315-319.

36. Li Renxian. Calculation of Loads on Crankshafts of Internal Combustion Engines Based on Continuous Beam Model [J]. Journal of Southwest Jiaotong University, 1998, 33(2): 164-169.

37. Li Shen, Zhao Shusen. Study on Substructure Method for Torsional Vibration of a Branch Shafting System [J]. Journal of Vibration and Shock, 2007-10-033.

38. Liang Xingyu, Shu Gequn, Li Donghai, Shen Yinggang. Torsional Vibration Analysis Based on Multi-body Dynamics of Flexible Crankshaft System [J]. Chinese Combustion Engine Engineering, in 2008, the fourth period.

39. Liang Xingyu, Shu Gequn, Li Donghai, Shen Yinggang. Torsional Vibration Analysis Based on Multi-body Dynamics of Flexible Crankshaft System [J]. Chinese Combution Engine Engineering, In 2008, the fourth period.

40. Lin Dayuan, Wang Xu, Song Xueren. Diagnosis of Crankshaft Cracks via Measurement of Crankshaft—Flywheel System Torsional Vibration Mode [J]. Transactions of Csice, 1989, 7(3):215-222.

41. Lin Ruilin, Huang Cihao, Tang Kaiyuan. The Incremental Harmonic Balance Method for CalculatingResponse of Non-Linear Torsional

Vibration of a Diesel Shafting [J].Transactions of Csice, 2002,20(2):185-187.

42. Lin Sen, Zhang Hongtian, Geng Ruiguang, et al. Study on numerical emulator of shafts' torsion vibration of internal-combustion engine based upon time step integration method [J]. Journal of Heilongjiang Institute of Technology, 2008, 22(1):1-7.

43. Liu Hui, Xiang Changle, Zheng Muqiao. Torsional Vibration on the Complex Shaft System of a Vehicular Powertrain [J]. Journal of Beijing Institute of Technology, 2002,22(6):699-703.

44. Liu Shengtian, Hydraulic dual-mass flywheel torsional vibration damper, [J]. Motor Transport,1998,9: 17-19.

45. Liu Shengtian. Double Mass Flywheel Type Torsional Vibration Damper [J]. Automobile Technology, 1997,1: 23-27.

46. Luo Zhouquan. Characters of Shafting Torsional Vibration and Analysis on Damping Measures of 6180 Series Marine Engine [J]. Design and Manufacture of Diesel Engine, 2000, 92 (3):3-7.

47. M Hosek, H Elmal, N Olgac. A Tunable Torsional Vibration Absorber: the Centrifugal Delayed Resonator[J]. Journal of Sound and Vibration, 1997,205(2):151-165.

48. N A Halliwell. The Laser Torsional Vibrometer: A Step Forward in Rotating Machinery Diagnostics[J]. Journal of Sound and Vibration, 1996,190(3):399-418.

49. Natarajan M, Frame E A, Naegeli D, et al. Oxygenates for ad vanced petroleum-based diesel fuels: part1. Screening and selection methodology for the oxygenates[C]. SAE 2001- 01-3631.

50. Nestorides E J. A Handbook on Torsional Vibration[M]. London: Cambridge University Press, 1958. Wilson W K. Practical Solution of Torsional Vibration Problem[M]. London: Campman and Hall, 1963.

51. Qu xiaoming, Gu yanhua, Wang xiaohu. Modeling and calculation of torsional vibration of internal combustion engine [J]. Internal combustion Engines, 2000,4:6-9.

52. Qu Zhihao, Chai Shaokuan, Ye Qianyuan. Analysis of Dynamic Characteristics and Chatter Of A 1420 Cold Tandem Rolling Mill [J]. Journal of Vibration and Shock, 2006, 25(4): 25-29.

53. S F Asokanthan, P A Meehans. Non-linear Vibration of a Torsional System Driven by a Hook's Joint[J]. Journal of Sound and Vibration, 1999,226(3):441-467.

54. Shen Hongbin Shu Gequn. An Investigation into Dynamic Measurement

of Longitudinal/Bend/Tortional Vibration in Shaft System of Internal Combustion Engines [J]. Small Internal Combustion Engine, 2000, 29(6):9-10.

55. Shen Tumiao. Dynamic Measurement and Calculation of Twisting Stress of Crankshaft and Flywheel System in Diesel Type TC 387 [J]. Transactions of Csice, l993, 11(6):243- 248.

56. Sheng Gang, Chen Zhiyan, Li Liangfeng. Torsional Vibration of the Diesel Crankshaft System with Variable Inertia [J]. Transactions of Csice, 1991,9(2):143-149.

57. Shu Gequn, Hao Zhiyong. An Investigation on the Torsional Elastic Wave Theory for the Calculation of Multi-Stepped Shaft Torsional Vibration Response [J]. China Mechanical Engineering, 1999,10(11):1277-1279.

58. Shu Gequn, Hao Zhiyong. Experiments on the Crankshaft Torsional/ Bending Vibration Damper in Internal Combustion Engine [J]. Journal of Tianjin University, 1997,30(6): 806-811.

59. Shu Gequn, Lü Xingcai, QIin De, Su Yanling. Experimental Study on Hybrid Damper of HighSpeed Diesel Engine Crankshaft [J]. Transactions of Csice, 2000,20(6): 537-540.

60. Shu Gequn, Lü Xingcai. Calculation Method for Continuous Distribution Model of Torsional Vibration of a Crankshaft [J]. Transactions of The Chinese Society of Agricultural Machinery, 2004,35 (4) 36-39.

61. Shu Gequn. Effect of Bending Vibration Damper on Properties of Torsional [J]. Transactions of Csice, 1998,16(3): 348-353.

62. Song Jingbo. Internal combustion Engines, l993, (3):38-40.

63. Song Xigeng, Song Tianxiang, Li Bozhong. On Coupled Vibrations of Reciprocating Engine Shaft Systems (Part 3)- Calculation Method of Coupled at Same and Double Frequencies [J]. Transactions of Csice, 1994,12(2):115-120.

64. T J Miless, M Lucast, N A Halliwell. Torsional and Bending Vibration Measurement on Rotors Using Laser Technology[J]. Journal of Sound and Vibration, 1999,226(3): 441-467.

65. Tan Daming. Vibration control of internal combustion engine [M]. Southwest Jiaotong university press, 1993.

66. Tong-Qun Han, Sheng-Hua Yao, Sheng-Jun Wu. Analysis of Torsional Vibration of Crankshaft System of a Vehicle Diesel Engine Based on EXCITE-designer Software. Journal of Hubei Automotive Industries Institute, 2005, 19(4): 5-8.

67. W J Hshen. on the Vibration Analysis of Multi-Branch Torsional

System[J]. Journal of Sound and Vibration,1999,224(2):209-220.
68. Wang Changmin, Zhu Dejun. Calculation of Torsional Stiffness of Engine Crankthrow by Finite Element Method [J]. Transactions of CSICE, 1991,9 (2) : 177-183.
69. Wang Guozhi, Modal Analysis Method in Research on Torsional Vibration of Diesel Engine Shafting System [J]. Transactions of Csice, l986,4(3):249-259.
70. Wang Ting, Cheng Peng. Development of Computer Testing System of Crankshaft Torsional Vibration [J]. Natural Science Journal of Jilin University of Technology, 201,21(3):74-77.
71. Wei Haijun. Modification of Some Formulas for Calculating Shaft Torsional Vibration [J]. Journal of Vibration and Shock, 2006, 25 (2): 166-168.
72. Wen Bangchun, Li Yinong, Han Qingkai, Analytical Approach and Application of Nonlinear Vibration Theory [M]. Shenyang: Northeast University Press, 2001.
73. Li Bozhong, Song Tianxiang, Song Xigeng. On Coupled Vibrations of Reciprocating Engine Shaft Systems(Part 1) Axial Vibration Caused by Torsional Vibration [J]. Transactions of Csice, 1989,7(1): 1-6. Li
74. Weng Wenhua. Silicone Oil Damper [J]. CCEC Science & Technology, 1997,1:19-23.
75. Xia Qingjie. 16V240ZJC Diesel Engine Torsional Vibration Measurement and Analysis[J]. Si Ji Science & Technology,1998,4:16-19.
76. Xiang Jianhua, Liao Ridong, Zhang Weizheng. Study and Application of Graphic Modeling Techniques of Shaft Torsional Vibration Based on System Matrix Method [J]. Acta Armamentarii, 2005-03-001
77. Xiang Jianhua, Liao Ridong, Zhang Weizheng. Study and Application of Graphic Modeling Techniques of Shaft Torsional Vibration Based on System Matrix Method [J]. Acta Armamentarii, 2005, 26(3): 294-298.
78. Xiang Shuhong, Qiu Jibao, Wang Dajun. The Resent Progresses on Modal Analysis and Dynamic Sub-Structure Methods [J]. Advances In Mechanics, 2004, 34(3): 289-303.
79. Xu Xintian. Measure of the torsional vibration control [J]. Wuhan Shipbullding, 1999, 1:17- 18.
80. Yan Jiabin. Engine Crankshaft Torsional Vibration Damper, World Rubbre Industry, 1998, 25(4):26-30.
81. Yang Yanfu, Measurement and Application of Torsional Vibration in Engine[J]. Ccec Science & Technology,1997,4:1-6.

82. Ying Qiguang, Bao Defu and Others. A New Fault-diagnosing Technique of Diesel Engine based on Shafting Vibration [J]. Ship Engineering, 1995, (4):33-35.
83. Ying Qiguang, Li Bozhong. A New Effective Method to Diagnose the Troubles in the Technical Condition of the Diesel Engine [J]. Journal of Shanghai Maritime University, 1990, (3):20-30.
84. Yu Qi. Torsional vibration of internal combustion engine [M].Beijing: National Defense Industry Press, 1985.
85. Yu Yinghui, Zhang Baohui. Development And Prospect of Research on Turbine-Generator Shaft Torsional Oscillation [J]. Automation Of Electric Power Systems, 1999,23(10):56-60.
86. Z P Mourelatos. An Efficient Crankshaft Dynamic Analysis Using Sub structuring with RITZ VECTORS[J]. Journal of Sound and Vibration, 2000,238(3):495-527.
87. Zhang Hongtian, Zhang Zhihua, LIU Zhigang. A Study on Coupled Axial and Torsional Vibration of Marine Propulsion Shafting [J]. Shipbuilding of China, 1995,129(2):68- 76.
88. Zhang Xiaoling, Tang Xikuan. Analysis of torsional vibration measuring errors and their correct method [J]. Journal Of Tsinghua University (Science And Technology), 1997, 37(8): 9-12.
89. Zhang Yong, Jiang Zikang. Analysis of the coupled flexural-torsional vibrations of rotary shaft systems [J]. Journal of Tsinghua University(Science And Technology), 2000,40(6):80-83.
90. Zhang Yong, Jiang Zikang. Mathematic model of coupled bending and torsional vibration of shaft systems [J]. Journal of Tsinghua University(Science And Technology), 1998,38(8):114-117.
91. Zhang Yong, Jiang Zikang. Numerical Analyses of the Coupling of the Flexural and Torsional Vibrations of Rotary Shaft System[J]. Turbine Technology,1999,41(5): 280- 283.
92. Zhang Zhihua, Tang Mi. The Numerical Computation on Torsional Vibration of Diesel Nonlinear Shafting, Transactions of Csice, 1987,5(4): 353-361.
93. Zhao Qian, Hao Zhiyong. Optimal Design of Torsional Damper of Diesel Engine Crankshaft System [J]. Transactions of the Chinese Society of Agricultural Machinery, 2000,31 (5): 91-93.
94. Zhou Shengfu, Li Cuncai. Function and Behaviour of Torsional Damper of Crankshaft [J]. Vehicle Engine, 1993,4:23-28.

95. Zhu Menghua. Direct-Inverse Fourier Transformation Technique for Responses of NonLinear Torsional Vibration of Diesel Shafting [J]. Transactions of Csice, 1992,10(1):47-52

Chapter 12

MAGNETIC LEVITATION TECHNIQUE FOR ACTIVE VIBRATION CONTROL

Md. Emdadul Hoque and Takeshi Mizuno

Saitama University Japan

INTRODUCTION

This chapter presents an application of zero-power controlled magnetic levitation for active vibration control. Vibration isolation are strongly required in the field of high-resolution measurement and micromanufacturing, for instance, in the submicron semiconductor chip manufacturing, scanning probe microscopy, holographic interferometry, cofocal optical imaging, etc. to obtain precise and repeatable results. The growing demand for tighter production tolerance and higher resolution leads to the stringent requirements in these research and industry environments. The microvibrations resulted from the tabletop and/or the ground vibration should be carefully eliminated from such sophisticated systems. The vibration control research has been advanced with passive and active techniques. Conventional passive technique uses spring and damper as isolator. They are widely used to support the investigated part to protect it from the severe ground vibration or from direct disturbance on the table by using soft and stiff suspensions, respectively (Haris & Piersol, 2002; Rivin, 2003). Soft suspensions can be used because they provide low resonance frequency of the isolation system and thus reduce the frequency band of vibration amplification. However, it leads to potential problem with static stability due to direct disturbance on the table, which can be solved by using stiff suspension. On the other hand, passive systems offer good high frequency vibration isolation with low isolator damping at the cost of vibration amplification at the fundamental resonance frequency. It can be solved by using high value of isolator damping. Therefore, the performance of passive isolators are limited, because various trade-offs are necessary when excitations with a wide frequency range are involved. Active control technique can be introduced to resolve these drawbacks.

Active control system has enhanced performances because it can adapt to changing environment (Fuller et al., 1997; Preumont, 2002; Karnopp, 1995). Although conventional active control system achieves high performance, it requires large amount of energy source to drive the actuators to produce active damping force (Benassi et al., 2004a & 2004b; Yoshioka et al., 2001; Preumont et al., 2002; Daley et al., 2006; Zhu et al., 2006; Sato & Trumper, 2002). Apart from this, most of the researches use high-performance sensors, such as servo-type accelerometer for detecting vibration signal, which are rather expensive. These are the difficulties to expand the application fields of active control technique.

The development and maintenance cost of vibration isolation system should be lowered in order to expand the application fields of active control. Considering the point of view, a vibration isolation system have been developed using an actively zero-power controlled magnetic levitation system (Hoque et al., 2006; Mizuno et al., 2007a; Hoque et al., 2010a). In the proposed system, eddy-current relative displacement sensors were used for displacement feedback. Moreover, the control current converges to zero for the zero-power control system. Therefore, the developed system becomes rather inexpensive than the conventional active systems.

An active zero-power controlled magnetic suspension is used in this chapter to realize negative stiffness by using a hybrid magnet consists of electromagnet and permanent magnets. Moreover, it can be noted that realizing negative stiffness can also be generalized by using linear actuator (voice coil motor) instead of hybrid magnet (Mizuno et al., 2007b). This control achieves the steady state in which the attractive force produced by the permanent magnets balances the weight of the suspended object, and the control current converges to zero. However, the conventional zero-power controller generates constant negative stiffness, which depends on the capacity of the permanent magnets. This is one of the bottlenecks in the field of application of zero-power control where the adjustment of stiffness is necessary. Therefore, this chapter will investigate on an improved zero-power controller that has capability to adjust negative stiffness. Apart from this, zero-power control has inherently nonlinear characteristics. However, compensation to zero-power control can solve such problems (Hoque et al., 2010b). Since there is no steady energy consumption for achieving stable levitation, it has been applied to space vehicles (Sabnis et al., 1975), to the magnetically levitated carrier system in clean rooms (Morishita et al., 1989) and to the vibration isolator (Mizuno et al., 2007a). Six-axis vibration isolation system can be developed as well using this technique (Hoque et al., 2010a).

In this chapter, an active vibration isolation system is developed using zero-power controlled magnetic levitation technology. The isolation system is fabricated by connecting a mechanical spring in series with a suspension of negative stiffness (see Section 4 for details). Middle tables are introduced in between the base and the isolation table.

In this context, the nomenclature on the vibration disturbances, compliance and transmissibility are discussed for better understanding. The underlying concept on vibration isolation using magnetic levitation technique, realization of zero-power, stiffness adjustment, nonlinear compensation of the maglev system are presented in detail. Some experimental results are presented for typical vibration isolation systems to demonstrate that the maglev technique can be implemented to develop vibration isolation system.

VIBRATION SUPPRESSION TERMINOLOGY

Vibration Disturbances

The vibration disturbance sources are categorized into two groups. One is direct disturbance or tabletop vibration and another is ground or floor vibration.

Direct disturbance is defined by the vibrations that applies to the tabletop and generates deflection or deformation of the system. Ground vibration is defined by the detrimental vibrations that transmit from floor to the system through the suspension. It is worth noting that zero or low compliance for tabletop vibration and low transmissibility (less than unity) are ideal for designing a vibration isolation system.

Almost in every environment, from laboratory to industry, vibrational disturbance sources are common. In modern research or application arena, it is certainly necessary to conduct experiments or make measurements in a vibration-free environment. Think about a industry or laboratory where a number of energy sources exist simultaneously. Consider the silicon wafer photolithography system, a principal equipment in the semiconductor manufacturing process. It has a stage which moves in steps and causes disturbance on the table. It supports electric motors, that generates periodic disturbance. The floor also holds some rotating machines. Moreover, earthquake, movement of employees with trolley transmit seismic disturbance to the stage. Assume a laboratory measurement table in another case. The table supports some machine tools, and change in load on the table is a common phenomena. In addition, air compressor, vacuum pump, oscilloscope and dynamic signal analyzer with cooling fan rest on the floor. Some more potential energy souces are elevator mechanisms, air conditioning, rail and road transport, heat pumps that contribute to the vibrational background noise and that are coupled to the foundations and floors of the surrounding buildings. All the above sources of vibrations affect the system either directly on the table or transmit from the floor.

Compliance

Compliance is defined as the ratio of the linear or angular displacement to the magnitude of the applied static or constant force. Moreover, in case of a varying dynamic force or vibration, it can be defined as the ratio of the excited vibrational amplitude in any form of angular or translational displacement to the magnitude of the forcing vibration. It is the most extensively used transfer function for the vibrational response of an isolation table. Any deflection of the isolation table is demonstrated by the change in relative position of the components mounted on the table surface. Hence, if the isolation system has virtually zero or lower compliance (infinite stiffness) values, by definition, it is a better-quality table because the deflection of the surface on which fabricated parts are mounted is reduced. Compliance is measured in units of displacement per unit force, i.e., meters/Newton (m/N) and used to measure deflection at different frequencies.

The deformation of a body or structure in response to external payloads or forces is a common problem in engineering fields. These external disturbance forces may be static or dynamic. The development of an isolation table is a good example of this problem where such static and dynamic forces may exist. A static laod, such as that caused by a large, concentrated mass loaded or unloaded on the table, can cause the table to deform. A dynamic force, such as the periodic disturbance of a rotating motor placed on top of the table, or vibration induced from the building into the isolation table through its mounting points, can cause the table to oscillate and deform.

Assume the simplest model of conventional mass-spring-damper system as shown in Fig. 1(a), to understand compliance with only one degree-of-freedom system. Consider that a single frequency sinusoidal vibration applied to the system. From Newton's laws, the general equation of motion is given by

$$m\ddot{x} + c\dot{x} + kx = F_0 \sin \omega t , \qquad (1)$$

where m : the mass of the isolated object, x : the displacement of the mass, c : the damping, k : the stiffness, F_0 : the maximum amplitude of the disturbance, ω : the rotational frequency of disturbance, and t : the time.

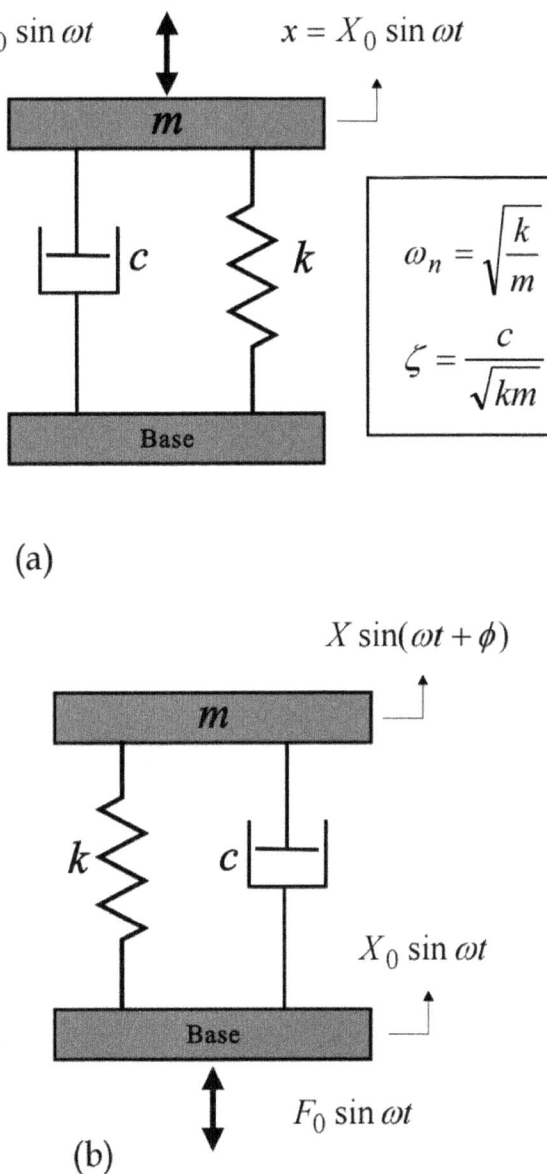

Figure. 1. Conventional mass-spring-damper vibration isolator under (a) direct disturbance (b) ground vibration.

The general expression for compliance of a system presented in Eq. (1) is given by

$$\text{Compliance} = \frac{x}{F} = \frac{1}{\sqrt{(k - m\omega^2)^2 + (c\omega)^2}}. \tag{2}$$

The compliance in Eq. (2) can be represented as

$$\text{Compliance} = \frac{x}{F} = \frac{1/k}{\sqrt{(1 - (\omega/\omega_n)^2)^2 + 4\zeta^2(\omega/\omega_n)^2}}, \tag{3}$$

where ω_n : the natural frequency of the system and z : the damping ratio.

Transmissibility

Transmissibility is defined as the ratio of the dynamic output to the dynamic input, or in other words, the ratio of the amplitude of the transmitted vibration (or transmitted force) to that of the forcing vibration (or exciting force).

Vibration isolation or elimination of a system is a two-part problem. As discussed in Section 2.1, the tabletop of an isolation system is designed to have zero or minimal response to a disturbing force or vibration. This is itself not sufficient to ensure a vibration free working surface. Typically, the entire table system is subjected continually to vibrational impulses from the laboratory floor. These vibrations may be caused by large machinery within the building as discussed in Section 2.1 or even by wind or traffic-excited building resonances or earthquake.

The model shown in Fig. 1(a) is modified by applying ground vibration, as shown in Fig. 1(b). The absolute transmissibility, T of the system, in terms of vibrational displacement, is given by

$$\frac{X}{X_0} = \sqrt{\frac{1 + 4\zeta^2(\omega/\omega_n)^2}{(1 - (\omega/\omega_n)^2)^2 + 4\zeta^2(\omega/\omega_n)^2}}. \tag{4a}$$

Similarly, the transmissibility can also be defined in terms of force. It can be defined as the ratio of the amplitude of force tranmitted (F) to the amplitude of exciting force (F0). Mathematically, the transmissibility in terms of force is given by

$$\frac{F}{F_0} = \sqrt{\frac{1+4\zeta^2(\omega/\omega_n)^2}{(1-(\omega/\omega_n)^2)^2 + 4\zeta^2(\omega/\omega_n)^2}}.$$ (4b)

Zero-Power Controlled Magnetic Levitation 3.1 Magnetic Suspension System Since last few decades, an active magnetic levitation has been a viable choice for many industrial machines and devices as a non-contact, lubrication-free support (Schweitzer et al., 1994; Kim & Lee, 2006; Schweitzer & Maslen, 2009). It has become an essential machine element from high-speed rotating machines to the development of precision vibration isolation system. Magnetic suspension can be achieved by using electromagnet and/or permanent magnet. Electromagnet or permanent magnet in the magnetic suspension system causes flux to circulate in a magnetic circuit, and magnetic fields can be generated by moving charges or current. The attractive force of an electromagnet, F can be expressed approximately as (Schweitzer et al., 1994)

$$F = K \frac{I^2}{\delta^2},$$ (5)

where K : attractive force coefficient for electromagnet, I : coil current, d : mean gap between electromagnet and the suspended object. Each variable is given by the sum of a fixed component, which determines its operating point and a variable component, such as

$$I = I_0 + i,$$ (6)

$$\delta = D_0 - x,$$ (7)

where I_0 : bias current, i : coil current in the electromagnet, D_0 : nominal gap, x : displacement of the suspended object from the equilibrium position.

Magnetic Suspension System with Hybrid Magnet

In order to reduce power consumption and continuous power supply, permanent magnets are employed in the suspension system to avoid providing bias current. The suspension system by using hybrid magnet, which consists of electromagnet and permanent magnet is shown in Fig. 2. The permanent magnet is used for the purpose of providing bias flux (Mizuno & Takemori, 2002). This control realizes the steady states in which the electromagnet coil current converges to zero and the attractive force produced by the permanent magnet balances the weight of the suspended object.

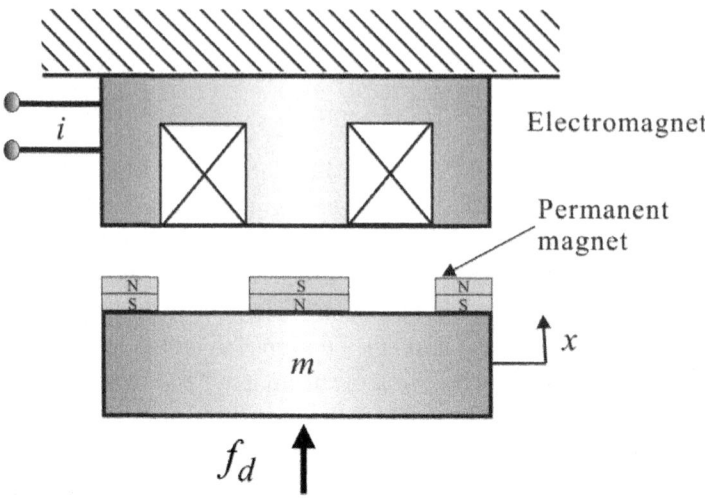

Figure. 2. Model of a zero-power controlled magnetic levitation.

It is assumed that the permanent magnet is modeled as a constant-current (bias current) and a constant-gap electromagnet in the magnetic circuit for simplification in the following analysis. Attractive force of the electromagnet, F can be written as

$$F = K\frac{(I_0 + i)^2}{(D_0 - x)^2},$$
(8)

where bias current, I_0 is modified to equivalent current in the steady state condition provided by the permanent magnet and nominal gap, D_0 is modified to the nominal air gap in the steady state condition including the height of the permanent magnet. Equation (8) can be transformed as

$$F = K\frac{I_0^2}{D_0^2}\left(1 - \frac{x}{D_0}\right)^{-2}\left(1 + \frac{i}{I_0}\right)^2.$$
(9)

Using Taylor principle, Eq. (9) can be expanded as

$$F = K\frac{I_0^2}{D_0^2}\left(1 + 2\frac{x}{D_0} + 3\frac{x^2}{D_0^2} + 4\frac{x^3}{D_0^3} + \ldots\right)\left(1 + 2\frac{i}{I_0} + \frac{i^2}{I_0^2}\right).$$
(10)

For zero-power control system, control current is very small, especially, in the phase approaches to steady-state condition and therefore, the higher-order terms are not considered. Equation (10) can then be written as

$$F = F_e + k_i i + k_s(x + p_2 x^2 + p_3 x^3 + \ldots), \quad (11)$$

where

$$F_e = K \frac{I_0^2}{D_0^2}, \quad (12)$$

$$k_i = 2K \frac{I_0}{D_0^2}, \quad (13)$$

$$k_s = 2K \frac{I_0^2}{D_0^3}, \quad (14)$$

$$p_2 = \frac{3}{2D_0}, \quad (15)$$

$$p_3 = \frac{4}{2D_0^2}. \quad (16)$$

For zero-power control system, the control current of the electromagnet is converged to zero to satisfy the following equilibrium condition

$$F_e = mg, \quad (17)$$

and the equation of motion of the suspension system can be written as

$$m\ddot{x} = F - mg. \quad (18)$$

From Eqs. (11), (17) and (18),

$$m\ddot{x} = k_i i + k_s(x + p_2 x^2 + p_3 x^3 + \ldots). \quad (19)$$

This is the fundamental equation for describing the motion of the suspended object.

Design of Zero-Power Controller

Negative stiffness is generated by actively controlled zero-power magnetic suspension. The basic model, controller and the characteristic of the zero-power control system is described below.

Model

A basic zero-power controller is designed for simplicity based on linearized equation of motions. It is assumed that the displacement of the suspended mass is very small and the nonlinear terms are neglected. Hence the linearized motion equation from Eq. (19) can be written as

$$m\ddot{x} = k_i i + k_s x . \tag{20}$$

The suspended object with mass of m is assumed to move only in the vertical translational direction as shown by Fig. 2. The equation of motion is given by

$$m\ddot{x} = k_s x + k_i i + f_d , \tag{21}$$

where x : displacement of the suspended object, k_s : gap-force coefficient of the hybrid magnet, k_i : current-force coefficient of the hybrid magnet, i : control current, f_d : disturbance acting on the suspended object. The coefficients k_s and k_i are positive. When each Laplace-transform variable is denoted by its capital, and the initial values are assumed to be zero for simplicity, the transfer function representation of the dynamics described by Eq. (21) becomes

$$X(s) = \frac{1}{s^2 - a_0}(b_0 I(s) + d_0 W(s)), \tag{22}$$

Where

$$a_0 = k_s / m, \; b_0 = k_i / m, \; \text{and} \; d_0 = 1/m.$$

Suspension with Negative Stiffness

Zero-power can be achieved either by feeding back the velocity of the suspended object or by introducing a minor feedback of the integral of current in the PD (proportional derivative) control system (Mizuno & Takemori, 2002). Since PD control is a fundamental control law in magnetic suspension, zero-power control is realized from PD control in this work using the second approach. In the current controlled magnetic suspension system, PD control can be represented as

$$I(s) = -(p_d + p_v s)X(s), \tag{23}$$

where p_d : proportional feedback gain, p_v : derivative feedback gain. Figure 3 shows the block diagram of a current-controlled zero-power controller where a minor integral feedback of current is added to the proportional feedback of displacement.

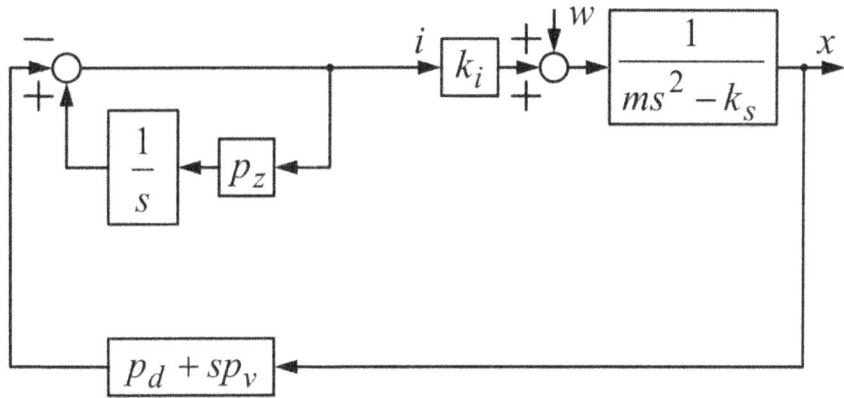

Figure. 3. Transfer function representation of the zero-power controller of the magnetic levitation system.

The control current of zero-power controller is given by

$$I(s) = -(\frac{s}{s-p_z}p_d + p_v s)X(s), \tag{24}$$

where p_z : integral feedback in the minor current loop. From Eqs. (22) to (24), it can be written as

$$\frac{X(s)}{W(s)} = \frac{(s-p_z)d_0}{s^3 + (b_0 p_v - p_z)s^2 + (b_0 p_d - b_0 p_v p_z - a_0)s + a_0 p_z}, \tag{25}$$

$$\frac{I(s)}{W(s)} = \frac{-s(sp_v + p_d - p_v p_z)d_0}{s^3 + (b_0 p_v - p_z)s^2 + (b_0 p_d - b_0 p_v p_z - a_0)s + a_0 p_z}. \tag{26}$$

To estimate the stiffness for direct disturbance, the direct disturbance, W(s) on the isolation table is considered to be stepwise, that is

$$W(s) = \frac{F_0}{s}, (F_0 : constant). \tag{27}$$

The steady displacement of the suspension, from Eqs. (25) and (27), is given by

$$\lim_{t \to \infty} x(t) = \lim_{s \to 0} sX(s) = -\frac{d_0}{a_0}F_0 = -\frac{F_0}{k_s}. \tag{28}$$

The negative sign in the right-hand side illustrates that the new equilibrium position is in the direction opposite to the applied force. It means that the system realizes negative stiffness. Assume that stiffness of any suspension is denoted by k. The stiffness of the zeropower controlled magnetic suspension is, therefore, negative and given by

$$k = -k_s. \tag{29}$$

Realization of Zero-Power

From Eqs. (26) and (27)

$$\lim_{t \to \infty} i(t) = \lim_{s \to 0} sI(s) = 0. \tag{30}$$

It indicates that control current, all the time, converges to zero in the zero-power control system for any load.

Stiffness Adjustment

The stiffness realized by zero-power control is constant, as shown in Eq. (29). However, it is necessary to adjust the stiffness of the magnetic levitation system in many applications, such as vibration isolation systems. There are two approaches to adjust stiffness of the zeropower control system. The first one is by adding a minor displacement feedback to the zeropower control current, and the other one is by adding a proportional feedback in the minor current feedback loop (Ishino et al., 2009). In this research, stiffness adjustment capability of zero-power control is realized by the first approach. Figure 4 shows the block diagram of the modified zero-power controller that is capable to adjust stiffness. The control current of the modified zero-power controller is given by

$$I'(s) = -(\frac{p_d s}{s - p_z} + \frac{p_v s^2}{s - p_z} + p_s)X(s), \tag{31}$$

where p_s : proportional displacement feedback gain across the zero-power controller.

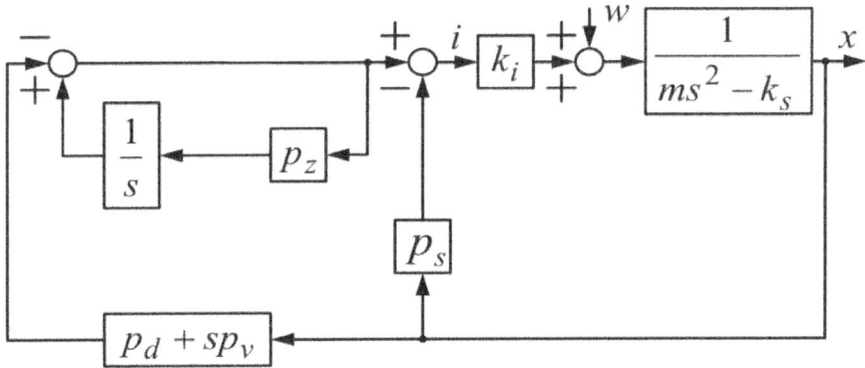

Figure. 4. Block diagram of the modified zero-power controller that can adjust stiffness.

The transfer-function representation of the dynamics shown in Fig. 4 is given by

$$\frac{X(s)}{W(s)} = \frac{(s-p_z)d_0}{s^3 + (b_0 p_v - p_z)s^2 + (b_0 p_d + b_0 p_s - a_0)s + a_0 p_z - b_0 p_s p_z}. \quad (32)$$

From Eqs. (27) and (32), the steady displacement becomes

$$\lim_{t \to \infty} x(t) = \lim_{s \to 0} sX(s) = -\frac{d_0 p_z}{a_0 p_z - b_0 p_s p_z} F_0 = -\frac{F_0}{k_s - k_i p_s} \quad (33)$$

Therefore, the stiffness of the modified system becomes

$$k = -k_s + k_i p_s. \quad (34)$$

It indicates that the stiffness can be increased or decreased by changing the feedback gain p_s.

Nonlinear Compensation of Zero-Power Controller

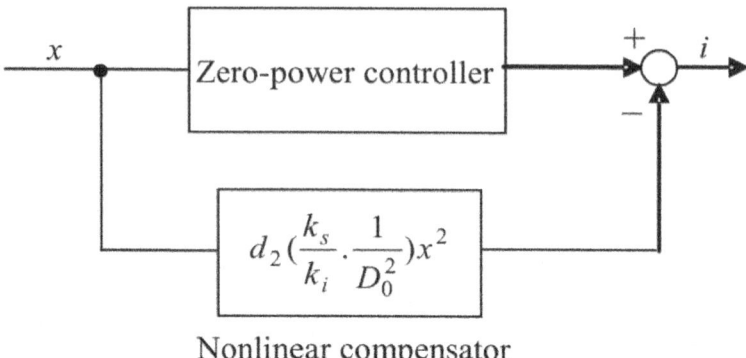

Figure. 5. Block diagram of the nonlinear compensator of the zero-power controlled magnetic levitation.

It is shown that the zero-power control can generate negative stiffness. The control current of the zero-power controlled magnetic suspension system is converged to zero for any added mass. To counterbalance the added force due to the mass, the stable position of the suspended object is changed. Due to the air gap change between permanent magnet and the object, the magnetic force is also changed, and hence, the negative stiffness generated by this system varies as well according to the gap (see Eq. (14)). To compensate the nonlinearity of the basic zero-power control system, the first nonlinear terms of Eq. (19) is considered and added to the basic system. From Eq. (19), the control current can be expressed as

$$i = i_{ZP} - d_2\left(\frac{k_s}{k_i} \cdot \frac{1}{D_0^2}\right)x^2 ,\tag{35}$$

where d_2 : the nonlinear control gain and, i_{zp} : the current in the zero-power controller, k_s, k_i and D_0 are constant for the system. The square of the displacement (x^2) is fed back to the normal zero-power controller. The block diagram of the nonlinear controller arrangement is shown in Fig. 5. The air gap between the permanent magnet and the suspended object can be changed in order to choose a suitable operating point. It is worth noting that the nonlinear compensator and the stiffness adjustment controller can be used simultaneously without instability. Moreover, performance of the nonlinear compensation could be improved furthermore if the second and third nonlinear terms and so on are considered together.

VIBRATION SUPPRESSION USING ZERO-POWER CONTROLLED MAGNETIC LEVITATION

Theory of Vibration Control

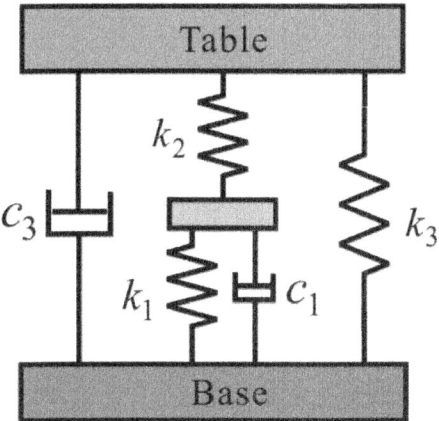

Figure. 6. A model of vibration isolator that can suppress both tabletop and ground vibrations.

The vibration isolation system is developed using magnetic levitation technique in such a way that it can behave as a suspension of virtually zero compliance or infinite stiffness for direct disturbing forces and a suspension with low stiffness for floor vibration. Infinite stiffness can be realized by connecting a mechanical spring in series with a magnetic spring that has negative stiffness (Mizuno, 2001; Mizuno et al., 2007a & Hoque et al., 2006). When two springs with spring constants of k_1 and k_2 are connected in series, the total stiffness k_c is given by

$$k_c = \frac{k_1 k_2}{k_1 + k_2}.$$
(36)

The above basic system has been modified by introducing a secondary suspension to avoid some limitations for system design and supporting heavy payloads (Mizuno, et al., 2007a & Hoque, et al., 2010a). The concept is demonstrated in Fig. 6. A passive suspension (k_3, k_3) is added in parallel with the serial connection of positive and negative springs. The total stiffness $\tilde{k}_c$ is given by

$$\tilde{k}_c = \frac{k_1 k_2}{k_1 + k_2} + k_3.$$
(37)

However, if one of the springs has negative stiffness that satisfies

$$k_1 = -k_2,$$
(38)

the resultant stiffness becomes infinite for both the case in Eqs. (36) and (37) for any finite value of k_3, that is

$$\left|\tilde{k}_c\right| = \infty.$$
(39)

Equation (39) shows that the system may have infinite stiffness against direct disturbance to the system. Therefore, the system in Fig. 6 shows virtually zero compliance when Eq. (38) is satisfied. On the other hand, if low stiffness of mechanical springs for system (k_1, k_3) are used, it can maintain good ground vibration isolation performance as well.

Typical Applications of Vibration Suppression

In this section, typical vibration isolation systems using zero-power controlled magnetic levitation are presented, which were developed based on the principle discussed in Eq. (37). The isolation system consists mainly of two suspensions with three platforms- base, middle table and isolation table. The lower suspension between base and middle table is of positive stiffness and the upper suspension between middle table and base is of negative stiffness realized by zero-power control. A passive suspension directly between base and isolation table acts as weight support mechanism.

A typical single-axis and a typical six-axis vibration isolation apparatuses are demonstrated in Fig. 7. The single-axis apparatus (Fig. 7(a)) consisted of a circular base, a circular middle table and a circular isolation table. The height, diameter and weight of the system were 300mm, 200mm and 20 kg, respectively. The positive stiffness in the lower part was realized by three mechanical springs and an electromagnet. To reduce coil current in the electromagnet, four permanent magnets (15mm×2mm) were used. The permanent magnets are made of Neodymium-Iron-Boron (NdFeB). The stiffness of each coil springs was 3.9 N/mm. The electromagnet coil had 180-turns and 1.3Ω resistance. The wire diameter of the coil was 0.6 mm. The relative displacement of the base to middle table was measured by an eddy-current displacement sensor, provided by Swiss-made Baumer electric. The negative stiffness suspension in the upper part was achieved by a hybrid magnet consisted of an electromagnet that was fixed to the middle table, and six permanent magnets attached to the electromagnet target on the isolation table. Another displacement sensor was used to measure the relative displacement between middle table to isolation table. The isolation table was also supported by three coil springs as weight support mechanism, and the stiffness of the each spring was 2.35 N/mm.

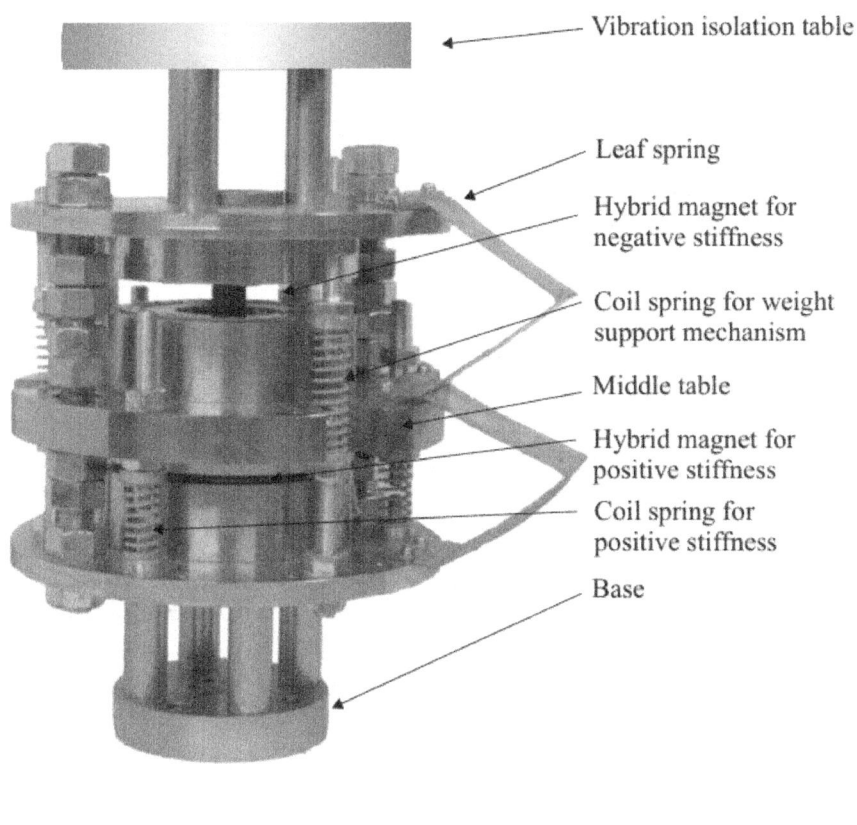

(a)

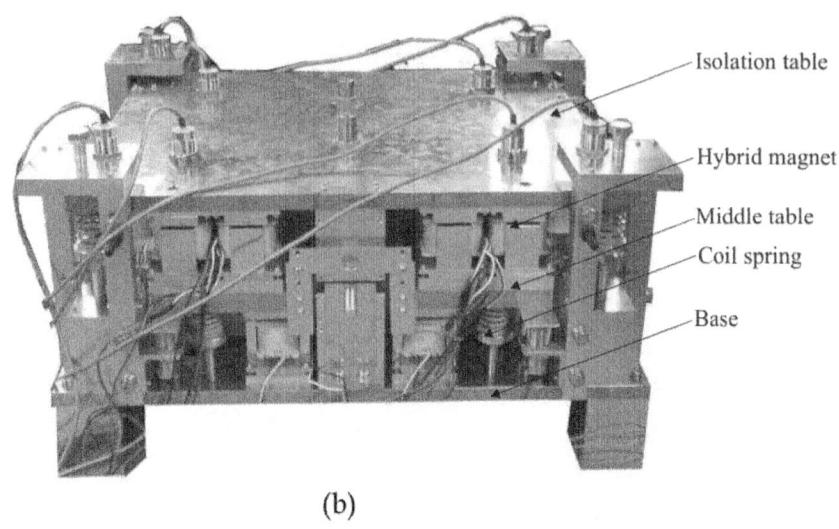

(b)

Figure. 7. Typical applications of zero-power controlled magnetic levitation for active vibration control (a) single-degree-of-freedom system (b) six-degree-of-freedom system.

The six-axis vibration isolation system with magnetic levitation technology is shown in Fig. 7(b) (Hoque, et al., 2010a). It consisted of a rectangular isolation table, a middle table and base. A positive stiffness suspension realized by electromagnet and normal springs was used between the base and the middle table. On the other hand, a negative stiffness suspension generated by hybrid magnets was used between the middle table and the isolation table. The height, length, width and mass of the apparatus were 300 mm, 740 mm, 590 mm and 400 kg, respectively. The isolation and middle tables weighed 88 kg and 158 kg, respectively. The isolation table had six-degree-of-freedom motions in the x, y, z, roll, pitch and yaw directions.

The base was equipped with four pairs of coil springs and electromagnets to support the middle table in the vertical direction and six pairs of coil springs and electromagnets (two pairs in the x-direction and four pairs in the y-direction) in the horizontal directions. The middle table was equipped with four sets of hybrid magnets to levitate and control the motions of the isolation table in the vertical direction and six sets of hybrid magnets (two sets in the x-direction and four sets in the y-direction) to control the motions of the table in the horizontal directions. The isolation table was also supported by four coil springs in the vertical direction and six coil springs (two in the x-direction and four in the y-direction) in the horizontal directions as weight support mechanism. Each set of hybrid magnet for zeropower suspension consisted of five square-shaped permanent magnets (20 mm×20 mm×2 mm) and five 585-turn electromagnets. The spring constant of each normal spring was 12.1 N/mm and that of weight support spring was 25.5 N/mm. There was flexibility to change the position of the weight support springs both in the vertical and horizontal directions to make it compatible for designing stable magnetic suspension system using zero-power control. The relative displacements of the isolation table to the middle table and those of the middle table to the base were detected by eight eddy-current displacement sensors attached to the corners of the isolation table and the base. A DSP-based digital controller (DS1103) was used for the implementation of the designed control algorithms by simulink in Matlab. The sampling rate was 10 kHz.

Experimental Demonstrations

Several experiments have been conducted to verify the aforesaid theoretical analysis. The nonlinear compensation of zero-power controlled magnetic levitation, stiffness adjustment of the levitation system are confirmed initially. Then the characteristics of the developed isolation systems are measured in terms of compliance and transmissibility.

Nonlinear Compensation of Magnetic Levitation System

First of all, zero-power control was realized between the isolation table and the middle table for stable levitation. Static characteristic of the zero-power controlled magnetic levitation was measured as shown in Fig. 8 when the payloads were increased to produce static direct disturbances on the table in the vertical direction. In this case, the middle table was fixed and the table was levitated by zero-power control. The result presents the load-stiffness characteristic of the zero-power control system. The figure without nonlinear compensation indicates that there was a wide variation of stiffness when the downward load force changed. For the uniform load increment, the change of gap was not equal due to the nonlinear magnetic force. Therefore, the negative stiffness generated from zero-power control was nonlinear which may severely affect the vibration isolation system.

To overcome the above problem, the nonlinear compensator was introduced in parallel with the zero-power control system. The nonlinear control gain (d_2) was chosen by trial and error method. The gap (D_0) between the table and the electromagnet was 5.1 mm after stable levitation by zero-power control. The value of k_s and k_1 were determined from the system characteristics. The load-stiffness characteristic using nonlinear compensation is also shown in the figure. It is obvious from the figure that the linearity error was reduced when control gain (d_2) was increased. For $d_2 = 55$, the linearity error was very low and the stiffness generated from the system was approximately constant. This result shows the potential to improve the static response performance of the isolation table to direct disturbance.

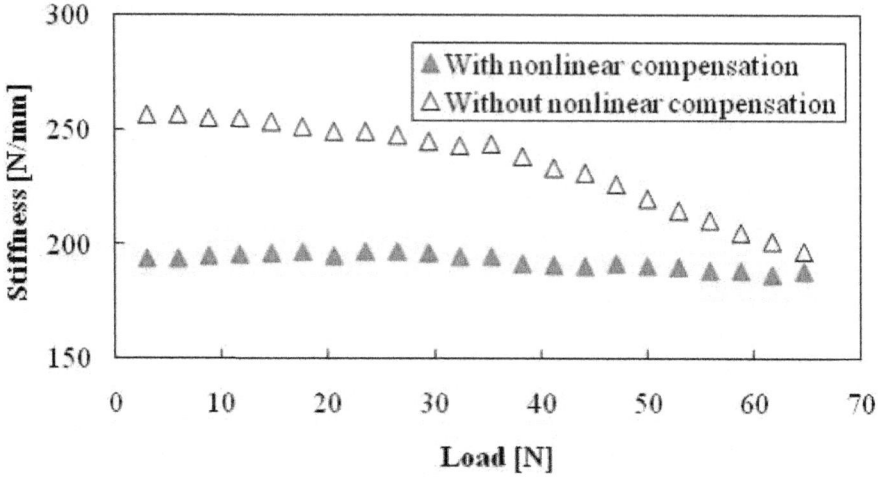

Figure. 8. Nonlinear compensation of the conventional zero-power controlled magnetic levitation system.

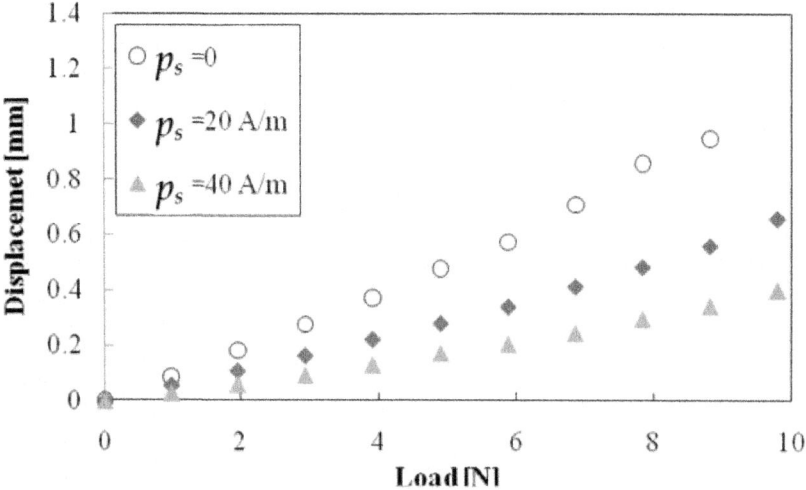

Figure. 9. Load-displacement characteristics of the modified zero-power controlled magnetic levitation system.

Stiffness Adjustment of Zero-Power Controlled Magnetic Levitation

The experiments have been carried out to measure the performances of the modified zeropower controller. Figure 9 shows the load-displacement characteristics of the system with the improved zero-power controller (Fig. 4). When the proportional feedback gain, $p_s = ,0$ it can be considered as a conventional zero-power controller (Fig. 3). The result shows that when the payloads were put on the suspended object, the table moved in the direction opposite to the applied load, and the gap was widened. It indicates that the zero-power control realized negative displacement, and hence its stiffness is negative, as described by Eqs. (28) and (29). The conventional zero-power controller ($p_s = 0$) realized fixed negative stiffness of magnitude -9.2 N/mm. When the proportional feedback gain, p_s was changed, the stiffness also gradually increased. When = 40 s p A/m, negative stiffness was increased to -21.5 N/mm. It confirms that proportional feedback gain, s p can change the stiffness of the zero-power controller, as explained in Eq. (34).

Experimental Results with Vibration Isolation System

Further experiments were conducted with the linearized zero-power controller with the vibration isolation system, as shown in Fig. 10. In this case, the positive and negative stiffness springs were, then, adjusted to satisfy Eq. (38). The stiffness could either be adjusted in the positive or negative stiffness part. In the former, PD control could be used in the electromagnets that were employed in parallel with the coil springs. The latter technique was presented in Section 4.3.2. For better performance, the latter was adopted in this work.

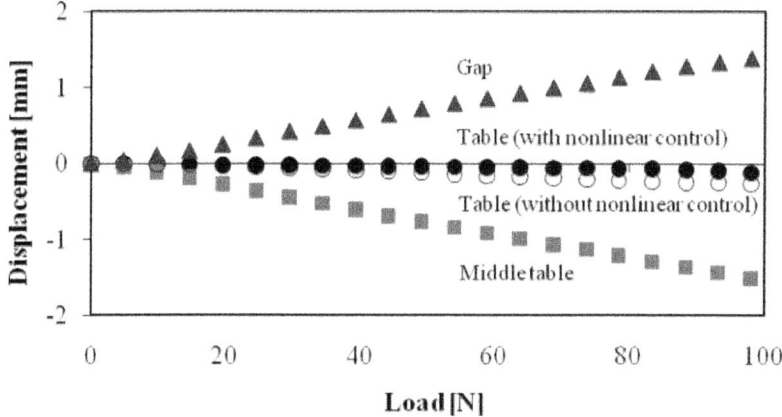

Figure. 10. Static characteristics of the isolation table with and without nonlinear control.

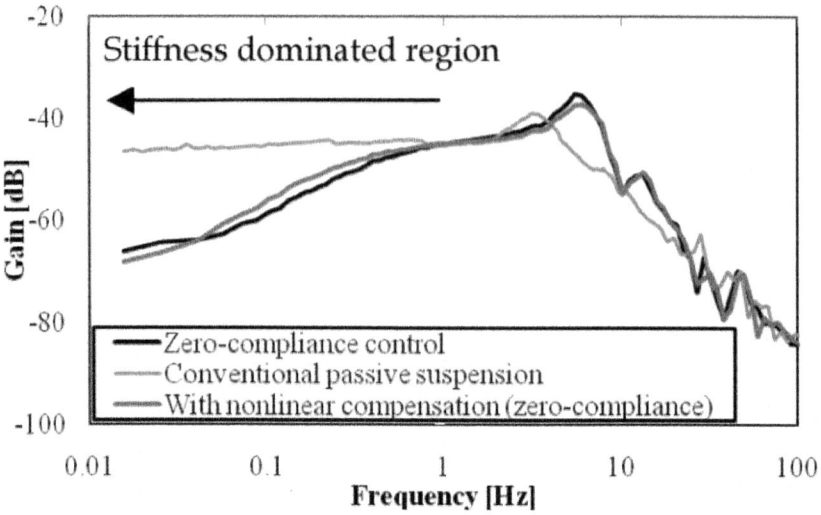

Figure. 11. Dynamic characteristics of the isolation table in the vertical direction.

Figure 10 demonstrates the performance improvement of the controller for static response to direct disturbance. The displacements of the isolation table and middle table were plotted against disturbing forces produced by payload in the vertical direction. It is clear that zero compliance to direct disturbance was realized up to 100 N payloads with nonlinear controller (d_2=55). The stiffness of the isolation system was increased to 960 N/mm which was approximately 2.8 times more than that of without nonlinear control. The figure illustrates significant improvement in rejecting on-board-generated disturbances.

The dynamic performance of the isolation table was measured in the vertical direction as shown in Fig. 11. In this case, the isolation table was excited to produce sinusoidal disturbance force by two voice coil motors which were attached to the base and can generate force in the Z-direction. The displacement of the table was measured by gap sensors and the data was captured by a dynamic signal analyzer. It is found from the figure that high stiffness, that means virtually zero-compliance, was realized at low frequency region (-66 dB[mm/N] at 0.015 Hz). It also demonstrates that direct disturbance rejection performance was not worsened even nonlinear zero-power control was introduced. Finally a comparative study of the disturbance suppression performance was conducted with zero-compliance control and conventional passive suspension technique as shown in the figure. The experiment was carried out with same lower suspension for ground vibration isolation. First, the isolation table was suspended by positive suspension (conventional spring-damper) and frequency response to direct disturbance was measured. The stiffness dominated region is marked in the figure, and it is seen from the figure that the displacement of the isolation table was almost same below 1 Hz (approximately -46 dB). However, when the isolation table was suspended by zero-compliance control satisfying Eqs. (38) and (39), displacement of the table was abruptly reduced at the low frequency region below 1 Hz (-66 dB at 0.015 Hz). It is confirmed from the figure that the developed zero-compliance system had better direct disturbance rejection performance over the conventional passive suspension even both the systems used similar vibration isolation performances.

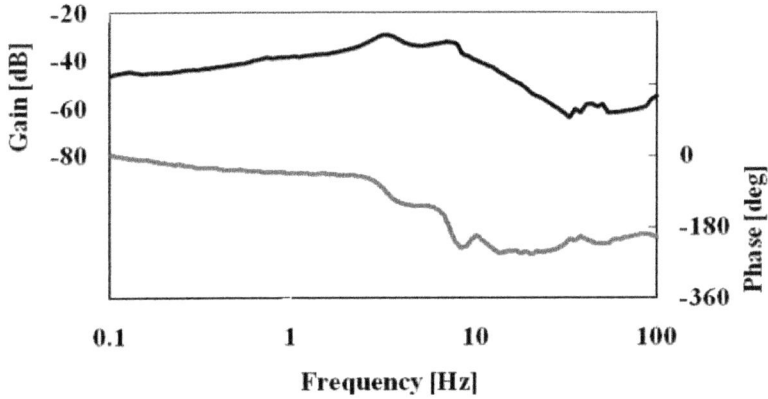

Figure. 12. Dynamic characteristics of the isolation table in the vertical direction.

The characteristics of the isolation table were further investigated by measuring the response of the table to direct disturbance in the horizontal directions as shown in Fig. 12. In this case, four voice coil motors were used to excite the isolation table along the horizontal direction. The results show the dynamic response of the isolation table when the table was excited along yaw mode. The response of the table to direct dynamic disturbance was captured by dynamic signal analyzer. The results justify that the displacements of the table to direct disturbance in the horizontal rotational motions were also low at the low frequency regions. The results confirmed that the isolation table was realized high stiffness against disturbing forces in the motion associated with horizontal direction.

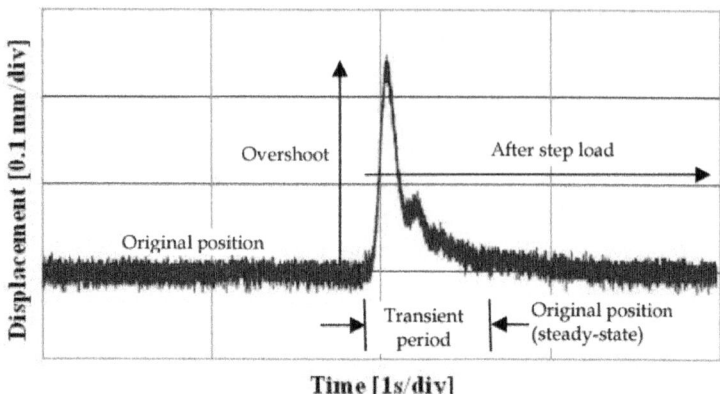

Figure. 13. Step response of the isolation table with magnetic levitation technology.

The step response of the isolation table is shown in Fig. 13. In this experiment, a stepwise disturbance was generated by suddenly removing a certain amount of load from the table and the response was measured. The results showed that the table moved upward in the direction of load removal and returned to the original position (steady-state) after certain period. However, there was a reverse action in case of step wise disturbance. Therefore, a peak was appeared due to the response of the step load. This unpleasant response might hamper the objective function of many advanced systems. It can be noted that a feedforward controller can be added in combination with zero-power control to overcome this problem.

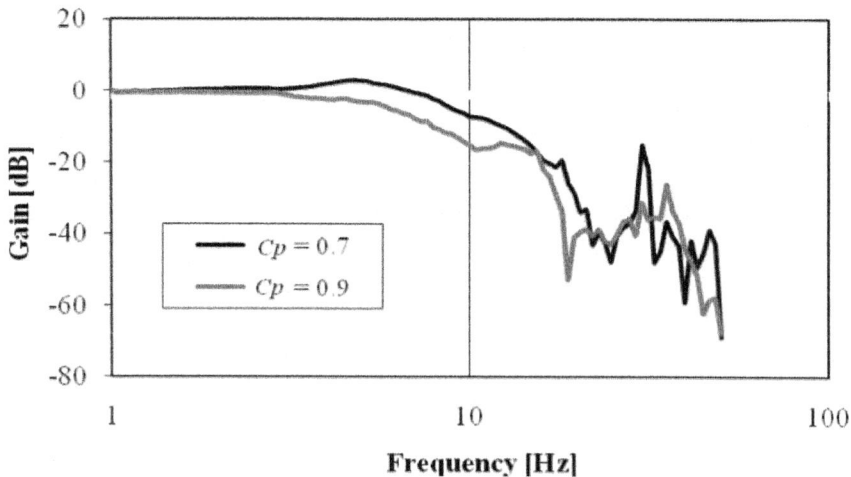

Figure. 14. Transmissibility characteristics of the isolation table.

Figure 14 shows the absolute transmissibility of the isolation table from the base of the developed system. In this case, the base of the system was sinusoidally excited in the vertical direction by a high-powered pneumatic actuator attached to the base, and the displacement transfer function (transmissibility) of the isolation table was measured from the base. The base displacement in the vertical direction was considered as input, and the output signal was the displacement of the isolation table. The damping coefficient (cp) between the base and the middle table played important role to suppress the resonance peak. The figure shows that the resonant peak was almost suppressed when cp was chosen as 0.9. It is clear from the figure that the developed system can effectively isolate the floor vibration that transmitted through the suspensions, such as active-passive positive suspensions and active zero-power controlled magnetic levitation.

CONCLUSIONS

A zero-power controlled magnetic levitation system has been presented in this chapter. The unique characteristic of the zero-power control system is that it can generate negative stiffness with zero control current in the steady-state which is realized in this chapter. The detail characteristics of the levitation system are investigated. Moreover, two major contributions, the stiffness adjustment and nonlinear compensation of the suspension system have been introduced elaborately. Often, there is a challenge for the vibration isolator designer to tackle both direct disturbance and ground vibration simultaneously with minimum system development and maintenance costs. Taking account of the point of view, typical applications of active vibration isolation using zero-power controlled magnetic levitation has been presented. The vibration isolation system is capable to suppress the effect of tabletop vibration as well as to isolate ground vibration. Some experimental demonstrations are presented that verifies the feasibility of its application in many industries and space related instruments. Moreover, it can be noted that a feedforward controller in combination with the zero-power controller can be used to improve the performance of the isolator to suppress direct disturbances.

ACKNOWLEDGMENT

The authors gratefully acknowledge the financial support made available from the Japan Society for the Promotion of Science as a Grant-in-Aid for scientific research (Grant no. 20.08380) for the foreign researchers and the Ministry of Education, Culture, Sports, Science and Technology of Japan, as a Grant-in-Aid for Scientific Research (B).

REFERENCES

1. Benassi, L. ; Elliot, S. J. & Gardonio, P. (2004a). Active vibration isolation using an inertial actuator with local force feedback control, Journal of Sound and Vibration, Vol. 276, No. 3, pp. 157-179
2. Benassi, L. & Elliot, S. J. (2004b). Active vibration isolation using an inertial actuator with local displacement feedback control, Journal of Sound and Vibration, Vol. 278, No. 4-5, pp. 705-724
3. Daley, S. ; Hatonen, J. & Owens, D. H. (2006). Active vibration isolation in a "smart spring" mount using a repetitive control approach, Control Engineering Practice, Vol. 14, pp. 991-997.
4. Fuller, C. R. ; Elliott, S. J. & Nelson, P. A. (1997). Active Control of Vibration, Academic Press, ISBN 0-12-269440-6, New York, USA
5. Harris, C. M. & Piersol, A. G. (2002). Shock and Vibration Handbook, McGraw Hill, Fifth Ed., ISBN 0-07-137081-1, New York, USA
6. Hoque, M. E. ; Takasaki, M. ; Ishino, Y. & Mizuno, T. (2006). Development of a three-axis active vibration isolator using zero-power control, IEEE/ASME Transactions on Mechatronics, Vol. 11, No. 4, pp. 462-470
7. Hoque, M. E. ; Mizuno, T. ; Ishino, Y. & Takasaki, M. (2010a), A six-axis hybrid vibration isolation system using active zero-power control supported by passive support mechanism, Journal of Sound and Vibration, Vol. 329, No. 17, pp. 3417-3430
8. Hoque, M. E. ; Mizuno, T. ; Kishita, D. ; Takasaki, M. & Ishino, Y. (2010b). Development of an Active Vibration Isolation System Using Linearized Zero-Power Control with Weight Support Springs, ASME Journal of Vibration and Acoustics, Vol. 132, No. 4, pp. 041006-1/9
9. Ishino, Y. ; Mizuno, T. & Takasaki, M. (2009). Stiffness Control of Magnetic Suspension by Local Feedback, Proceedings of the European Control Conference 2009, pp. 3881- 3886, Budapest, Hungary, 23-26 August, 2009
10. Karnopp, D. (1995). Active and semi-active vibration isolation, ASME Journal of Mechanical Design, Vol. 117, pp. 177-185
11. Kim, H. Y. & Lee, C. W. (2006). Design and control of Active Magnetic Bearing System With Lorentz Force-Type Axial Actuator, Mechatronics, vol. 16, pp. 13–20
12. Mizuno, T. (2001). Proposal of a Vibration Isolation System Using Zero-Power Magnetic Suspension, Proceedings of the Asia Pacific Vibration Conference 2001, pp. 423-427, Hangzhau, China
13. Mizuno, T. & Takemori, Y. (2002). A transfer-function approach to the

analysis and design of zero-power controllers for magnetic suspension system, Electrical Engineering in Japan, Vol. 141, No. 2, pp. 933-940

14. Mizuno, T. ; Takasaki, M. ; Kishita, D. & Hirakawa, K. (2007a). Vibration isolation system combining zero-power magnetic suspension with springs, Control Engineering Practice, Vol. 15, No. 2, pp. 187-196

15. Mizuno, T. ; Furushima, T. ; Ishino, Y. & Takasaki, M. (2007b). General Forms of Controller Realizing Negative Stiffness, Proceedings of the SICE Annual Conference 2007, pp. 2995-3000, Kagawa University, Japan, 17-20 September, 2007

16. Morishita, M. ; Azukizawa, T. ; Kanda, S. ; Tamura, N. & Yokoyama, T. (1989). A new maglev system for magnetically levitated carrier system, IEEE Transaction on Vehicular Technology, Vol. 38, No. 4, pp. 230-236

17. Platus, D. L. (1991). Negative-stiffness-mechanism vibration isolation system, Proceedings of the SPIE, Vibration Control in Microelectronics, Optics, and Metrology, Vol. 1619, pp. 44-54

18. Preumont, A. (2002). Vibration Control of Active Structures, An Introduction, Kluwer, Second ed., ISBN 1-4020-0496-6, Dordrecht

19. Preumont, A. ; Francois, A. ; Bossens, F. & Hanieh, A. A. (2002). Force feedback versus acceleration feedback in active vibration isolation, Journal of Sound and Vibration, Vol. 257, No. 4, pp. 605-613

20. Rivin, E. I. (2003). Passive Vibration Isolation, ASME Press, ISBN: 0-7918-0187-X, New York, USA

21. Sabnis, A. V. ; Dendy, J. B. & Schmitt, F. M. (1975). Magnetically suspended large momentum wheel, Journal of Spacecraft and Rockets, Vol. 12, pp. 420-427

22. Sato, T. & Trumper, D. L. (2002). A novel single degree-of-freedom active vibration isolation system, Proceedings of the 8th International Symposium on Magnetic Bearing, pp. 193-198, Japan, August 26-28, 2002

23. Schweitzer, G. ; Bleuler, H. & Traxler, A. (1994). Active Magnetic Bearings, vdf Hochschulverlag AG an der ETH Zurich, Zurich, Switzerlannd Schweitzer, G. & Maslen, E. H. (2009). Magnetic Bearings-Theory, Design, and Application to Rotating Machinery, ISBN : 978-3-642-00496-4, Springer, Germany

24. Yoshioka, H. ; Takahashi, Y. ; Katayama, K. ; Imazawa, T. & Murai, N. (2001). An active microvibration isolation system for hi-tech manufacturing facilities, ASME Journal of Vibration and Acoustics, Vol. 123, pp. 269-275

25. Zhu, W. H. ; Tryggvason, B. & Piedboeuf, J. C. (2006). On active acceleration control of vibration isolation systems, Control Engineering Practice, Vol. 14, No. 8, pp. 863-873.

CITATION

CHAPTER 1
Q. Ji, X. Ji, L. Ji and Y. Zheng, "A New Differential Operator Method to Study the Mechanical Vibration," Modern Mechanical Engineering, Vol. 2 No. 3, 2012, pp. 65-70. doi: 10.4236/mme.2012.23009.

CHAPTER 2
Mehmet Emin Yüksekkaya (2010). Characteristics of Mechanical Noise during Motion Control Applications, Motion Control, Federico Casolo (Ed.), ISBN: 978-953-7619-55-8, InTech, DOI: 10.5772/6974.

CHAPTER 3
Caiyou Zhao and Ping Wang, "Theoretical Modelling and Effectiveness Study of Slotted Stand-Off Layer Damping Treatment for Rail Vibration and Noise Control," Shock and Vibration, vol. 2015, Article ID 716382, 12 pages, 2015. doi:10.1155/2015/716382.

CHAPTER 4
Ye Lei, Jie Pan, and Meiping Sheng, "Study on Noise Prediction and Reduction in Coupled Workshops Using SEA Method," ISRN Mechanical Engineering, vol. 2011, Article ID 984253, 8 pages, 2011. doi:10.5402/2011/984253.

CHAPTER 5
Syed Md. Ihsanul Karim, Mohammad Asaduzzaman Chowdhury, and Md. Maksud Helali, "Influence of Sound Vibration on Diamond-Like Carbon Deposition Rate," ISRN Mechanical Engineering, vol. 2012, Article ID 676751, 8 pages, 2012. doi:10.5402/2012/676751.

CHAPTER 6

Zhu L, Ji S, Shen Q, Liu Y, Li J, Liu H (2013) A Noise Level Prediction Method Based on Electro-Mechanical Frequency Response Function for Capacitors. PLoS ONE 8(12): e81651. doi:10.1371/journal.pone.0081651.

CHAPTER 7

S. H. Gawande and S. N. Shaikh, "Experimental Investigations of Noise Control in Planetary Gear Set by Phasing," Journal of Engineering, vol. 2014, Article ID 857462, 11 pages, 2014. doi:10.1155/2014/857462.

CHAPTER 8

Bing Yang and Yan Liu, "Noise Source Identification of a Ring-Plate Cycloid Reducer Based on Coherence Analysis," Mathematical Problems in Engineering, vol. 2013, Article ID 875929, 5 pages, 2013. doi:10.1155/2013/875929.

CHAPTER 9

Marcel Janda, Ondrej Vitek and Vitezslav Hajek (2012). Noise of Induction Machines, Induction Motors - Modelling and Control, Prof. Rui Esteves Araújo (Ed.), ISBN: 978-953-51-0843-6, InTech, DOI: 10.5772/38152.

CHAPTER 10

Ferdinand Svaricek, Christian Bohn, Peter Marienfeld, Hans-Jürgen Karkosch and Tobias Fueger (2010). Automotive Applications of Active Vibration Control, Vibration Control, MickaÃƒÂ«l Lallart (Ed.), ISBN: 978-953-307-117-6, InTech, DOI: 10.5772/10149.

CHAPTER 11

Liang Xingyu, Shu Gequn, Dong Lihui, Wang Bin and Yang Kang (2011). Progress and Recent Trends in the Torsional Vibration of Internal Combustion Engine, Advances in Vibration Analysis Research, Dr. Farzad Ebrahimi (Ed.), ISBN: 978-953-307-209-8, InTech, DOI: 10.5772/16222.

CHAPTER 12

Emdadul Hoque and Takeshi Mizuno (2010). Magnetic Levitation Technique for Active Vibration Control, Magnetic Bearings, Theory and Applications, Bostjan Polajzer (Ed.), ISBN: 978-953-307-148-0, InTech.

INDEX

A

Acoustic Pressure 158
active noise and vibration control techniques (ANC/AVC) 182
antiresonance 40
automotive 118, 121, 142
automotive transmission 121

B

Barometric pressure 158

C

capacitor 95, 96, 97, 98, 101, 102, 103, 104, 105, 107, 108, 109, 110, 111, 113, 114, 115, 116
Chemical vapor deposition (CVD) 78
complex Fourier transform (CFT) 170
Compliance 246
Continuous mass model 204
Coupler shaft 126, 127
coupling loss factor (CLF) 60
Coupling Vibration Analysis 219
Crankshaft 203, 233, 235, 236, 238, 239, 240

D

Damping loss factor (DLF) 66
data acquisition (DAQ) 11
displacement 244, 246, 248, 249, 252, 253, 254, 255, 256, 258, 261, 262, 263, 264, 267, 268

E

eigenmodes 42
electrodynamic actuator 185
electro-mechanical frequency response function (EMFRF) 95, 96, 98, 114

F

Fast Fourier Transformation 170, 173, 175
frequency 27, 28, 29, 30, 33, 35, 40, 42, 43, 46, 47, 50, 52, 54
frequency spectrum diagram 147, 151
FxLMS algorithm 186, 193, 194

H

hydromount configuration 185
hyperparaboloid 187

I

in high-voltage direct-current (HVDC) 95
inverse real Fourier transform (IRFT) 171

L

laser Doppler velocimetry 216
localized electrochemical deposition (LECD) 90
Lovejoy Coupling 125

M

Magnetostriction 162
metallurgical equipments 2
microvibrations 243
multicrystal layer 78

N

nonresonant vibration 60

O

Oscilloscope 101, 109

P

PD (proportional derivative) 252
planetary gear 117, 118, 121, 122, 123, 125, 126, 129, 132, 141, 142, 143
portable digital vibrometer (PDV) 101

R

radiation 27, 28, 45, 52
resonance 40

S

Shunt 101, 116
Simple mass - spring model 204
sound intensity (SIL) 158
Sound Level Meter 170
sound power level (SWL) 158
sound pressure level (SPL) 158
sound vibration 77, 78, 80, 83, 85, 87, 88, 89, 90
statistical energy analysis (SEA) 57

T

torque 145
Torque Amplification Factor (TAF) 8
transmissibility 21
Transmissibility 248, 266

V

Vibration Disturbances 245
Vibration isolation 243, 248, 269
vibroacoustic system 59

X

X-ray diffraction (XRD) 81
X-ray spectrometry (EDX) 81